ROUTLEDGE LI
L(

Volume 8

TRUTH-FUNCTIONAL LOGIC

TRUTH-FUNCTIONAL LOGIC

J. A. FARIS

Routledge
Taylor & Francis Group

LONDON AND NEW YORK

First published in 1962 by Routledge & Kegan Paul Ltd

This edition first published in 2020
by Routledge
2 Park Square, Milton Park, Abingdon, Oxon OX14 4RN

and by Routledge
52 Vanderbilt Avenue, New York, NY 10017

Routledge is an imprint of the Taylor & Francis Group, an informa business

British Library Cataloguing in Publication Data
A catalogue record for this book is available from the British Library

ISBN: 978-0-367-41707-9 (Set)
ISBN: 978-0-367-81582-0 (Set) (ebk)
ISBN: 978-0-367-42017-8 (Volume 8) (hbk)
ISBN: 978-0-367-42612-5 (Volume 8) (pbk)
ISBN: 978-0-367-85398-3 (Volume 8) (ebk)

Publisher's Note
The publisher has gone to great lengths to ensure the quality of this reprint but points out that some imperfections in the original copies may be apparent.

Disclaimer
The publisher has made every effort to trace copyright holders and would welcome correspondence from those they have been unable to trace.

TRUTH-FUNCTIONAL LOGIC

BY

J. A. Faris

ROUTLEDGE & KEGAN PAUL

LONDON

First published 1962
by Routledge & Kegan Paul Ltd
Broadway House, 68–74 Carter Lane
London, E.C.4

Printed in Great Britain
by Latimer, Trend & Co. Ltd., Plymouth

CONTENTS

Contents

Chapter One

INTRODUCTION

1. Sentences, propositions, arguments and the point of view of logic. Logic begins as the study from a particular point of view of certain types of argument. A person who makes an inference, i.e. does a piece of reasoning, and expresses it in some way, whether privately or publicly, may be said to use an argument. An argument has two essential parts: a set of one or more *premisses* and a *conclusion*, and it is said to be *from* its premisses *to* its conclusion. In this book we shall take it that the basic elements of an argument, i.e. the premisses and the conclusion, are *propositions*. We begin by saying something about how this term is to be understood. Consider the following sentence:

The population of the world is increasing.

This sentence may be used to make a statement or express a proposition: namely the proposition that the population of the world is increasing. Nevertheless although the proposition is expressed by the sentence it is not identical with the sentence. For the sentence could be translated into another language: if this were done we should have a different sentence but one which expressed the same proposition.

Propositions have two important properties. One of these has already emerged: a proposition is expressed by an indicative sentence but is not identical with a sentence, and any particular proposition may be expressed equally well by any one of a number of different sentences. The

1

other property is that every proposition is either true or false.

The distinction between proposition and sentence, though of theoretical importance, will not be prominent in this book. On the contrary we will adopt for the sake of convenience two practices which will confine it to the background: in the first place in our illustrations we will normally use the same sentence to express the same proposition; in the second place when we wish to refer to a proposition which we express by means of a sentence *S* instead of using the phrase:

the proposition expressed by the sentence *S*

we shall say simply

the proposition *S*.

The second of these phrases is to be understood as an abbreviation for the first.

The other property we have referred to will, however, be constantly before us. It should be mentioned perhaps that it is sometimes questioned whether the property of being true or false is one that is in fact possessed by every proposition: it is suggested that there may be some propositions which are neither true nor false. However, although systems of logic, and indeed systems of truth-functional logic, have been devised which might be applicable if this were so, these must be taken to be outside our present field of study. It is a fundamental postulate of the system of logic which we are to expound that every proposition to which it applies is either true or false. We therefore regard this property of being either true or false as an essential property of propositions as we are using the term.

Let us now have an example of an argument. The following passage occurs in one of Bishop Berkeley's dialogues.[1]

[1] *First Dialogue between Hylas and Philonous.*

Introduction

'Because intense heat is nothing else but a particular kind of painful sensation; and pain cannot exist but in a perceiving being; it follows that no intense heat can really exist in an unperceiving corporeal substance.'

Berkeley is here presenting an argument which we may regard as having two premisses, namely the propositions:

Intense heat is nothing else but a particular kind of painful sensation,

and

Pain cannot exist but in a perceiving being.

The conclusion is the proposition:

No intense heat can really exist in an unperceiving corporeal substance.

In any argument the set of premisses is put forward as a reason for accepting the conclusion and in any presentation of an argument there is some indication of this relationship which enables us to identify premisses and conclusion respectively. In the present example the word *because* marks the premisses and the expression *it follows that* the conclusion. The word *therefore* is of course often used to mark the conclusion of an argument. The conclusion immediately follows the *therefore*; any propositions immediately preceding it are premisses. We will henceforward use this as our standard method of distinguishing premisses and conclusion. Premisses and conclusions will usually be numbered for ease of reference and an argument will sometimes be labelled with a capital letter and a number. This is exemplified in the following argument, which we call A1, in which the premisses are numbered (1) and (2) and the conclusion is numbered (3). It is obvious that A1 is very closely related to Berkeley's argument, though it might perhaps be disputed whether the two are identical:

Introduction

(A1) (1) Every instance of intense heat is a painful
          sensation.
      (2) No painful sensation is a thing capable of
          existing in an unperceiving being.
      Therefore (3) No instance of intense heat is a
          thing capable of existing in an unperceiv-
          ing being.

We began by saying that initially logic is the study of
arguments from a certain point of view. We must now
explain what this point of view is. Let us consider the
argument A1 which is set out above. A1 is a philosophical
argument; that is to say, it is of interest primarily to
philosophers. If we study this argument as philosophers
our interest will be to decide whether it provides us with
sufficient reason for believing in its conclusion. To answer
this question affirmatively we must convince ourselves that
two conditions are satisfied: (i) that the premisses are both
true; (ii) that the conclusion follows from the premisses.
The argument provides us with good reason for accepting
its conclusion as true if, but only if, it satisfies both conditions. The philosopher then will be interested in the two
questions: whether condition (i) is satisfied and whether
condition (ii) is satisfied. The logician as such, on the
other hand, in studying this argument is interested in the
second of these questions only, the question, that is, of
whether or not the conclusion follows from the premisses.
Another way of saying the same thing is to say that the
logician is interested, not in the question of whether the
conclusion of an argument is true, but rather in the question of whether the conclusion *must* be true *if* the premisses
are. Yet another way is to say that the logician is interested
not in whether the conclusion is true but in whether the
argument is *valid*. These are various alternative ways of
indicating roughly the point of view of the logician as
such. In the last of them we have used the term *valid*. The

concept of validity is of fundamental importance in the study of logic and we must now attempt to give a systematic account of it. To do this we must first introduce and explain the notion of a form of argument.

2. Argument forms; validity. Let us look again at the argument A1 and compare with it another argument which we shall call A2.

(A2) (4) Every visitor is a person now present.
      (5) No person now present is a prizewinner.
      Therefore (6) No visitor is a prizewinner.

These two arguments are obviously different from one another in at least one respect. They are different in respect of what they are about. A1 is about intense heat, painful sensations and things capable of existing in an unperceiving being; A2 is about visitors, persons now present, and prizewinners. Yet in another respect they are similar to one another. The respect in which A1 and A2 are similar is that they have the same form. The form of these two arguments may be expressed as follows and called *A*.

(A) Every [a   ] is a [b   ].
     No [b   ] is a [c   ].
     Therefore No [a   ] is a [c   ].

What is meant by saying that A1 and A2 have the same form A is this: if in the blank spaces in A marked *a*, *b* and *c* we write respectively:

intense heat, painful sensations, things capable of existing in an unperceiving being,

then we have the argument A1; whereas, on the other hand, if we write:

visitor, person now present, prizewinner,

then we have the argument A2.

Introduction

If two arguments are of the same form they may be said to *exemplify* that form or to be *exemplifications* of it. We should regard each form as having a set of exemplification rules which tell us what may and what may not be put in the blank spaces or gaps. We need to know this in order to know what is to count as an exemplification of a given form. In each case two rules at least are required. One rule, which we may call the *type rule*, governs the type of expression which is to be inserted in the blank spaces; the other rule, here called the *distribution rule*, governs the way in which expressions of the appropriate type or types may be distributed over the different spaces. For the form A the two rules are:

(Type rule)   Only a general term, i.e. a general noun or nominal phrase may be put in any space.

(Distribution rule)   Where two spaces are marked by the same letter they must be filled in in the same way; spaces marked by different letters may be filled in either in the same way or in different ways.

There is a generally accepted convention that this distribution rule which has just been stated for A applies to all forms. Hence a distribution rule is rarely stated explicitly. The type rule is often conveyed in the guise of a description. The space-labelling letters (*a*, *b*, *c* in our example) are known as *variables*; when we fill in a space we are said to be making a substitution for or replacing a variable. Sometimes a writer, after setting out a form which contains variables x, y and z, may say: 'where x, y and z are general term variables' or 'where x, y, z are proper name variables' or again 'where x, y and z are general terms' or 'where x, y and z are proper names'. Any such phrase really serves the purpose of giving the type rule for the form in question. Very often however the type rule is omitted altogether. There is a general principle that an

6

Introduction

argument form must be filled in in such a way that the result has for premises and conclusion significant propositions expressed by properly constructed sentences. With many forms this principle alone is sufficient to determine what sort of insertion is legitimate; apart from this it is often obvious from the context what type rule is intended.

The form A would more usually be written:

(A) (1) Every *a* is a *b*.
    (2) No *b* is a *c*.
    Therefore (3) No *a* is a *c*.

An argument is not an exemplification of a form if it cannot be obtained when proper substitutions are made for the variables in the form. For example the following argument is not an exemplification of the form A:

(B1) (1) Every visitor is a competitor in the last race.
    Therefore (2) Every competitor in the last race is a visitor.

To see that this is so it is sufficient to notice that the conclusion of every exemplification of A must begin with the word *No*, whereas the conclusion of B1 begins with the word *Every*.

B1 exemplifies the argument form:
(B) (1) Every *a* is a *b*.
    Therefore (2) Every *b* is an *a*.

For B as for A the type rule is that variables are to be replaced by general terms. We now set out a number of forms which have a common type rule entirely different from the one which applies to A and B.

(C) (1) p before q.
    (2) q before r.
    Therefore (3) p before r.

(D) (1) Aristotle said that p.
Therefore (2) p.

(E) (1) p and q.
Therefore (2) p.

(F) (1) p or q.
(2) It is not the case that p.
Therefore (3) q.

(G) (1) p.
Therefore (2) p and q.

For all these forms the type rule is that the letters *p*, *q* and *r* which are used as variables are to be replaced by propositions; to put it in another way, these letters are propositional variables. Thus if in C we replace *p* by *Hitler occupied the Rhineland*, *q* by *Mussolini invaded Abyssinia* and *r* by *The Second World War began* we obtain the argument:

(C1) (1) Hitler occupied the Rhineland before Mussolini invaded Abyssinia.
(2) Mussolini invaded Abyssinia before the Second World War began.
Therefore (3) Hitler occupied the Rhineland before the Second World War began.

Again, if in F we replace *p* by *James won the race* and *q* by *John competed*, we obtain the argument:

(F1) (1) James won the race or John competed.
(2) It is not the case that James won the race.
Therefore (3) John competed.

A form of argument may be either valid or invalid. A form of argument is said to be *valid* if there is no possible exemplification in which all the premisses are true but the conclusion is false; it is invalid if in at least one possible exemplification all the premisses are true but the conclusion is false. The argument forms A, C, E and F are all valid. But the following argument is an exemplification of B in which the sole premiss is true but the conclusion is false:

(B2) Every mouse is an animal.
      Therefore, Every animal is a mouse.

Hence the argument form B is invalid. Exemplifications of the forms D and G can also be found without difficulty which show that these forms too are invalid.

To say that an argument, e.g. A2, is valid is not to say the same thing as to say that an argument form, e.g. A, is valid. A definition of *valid argument* may now be given which makes use of the explanation already given of the meaning of valid argument form:

An argument is valid if it is an exemplification of at least one valid argument form; otherwise it is invalid.

In view of this definition it can be seen that when we say that an argument is valid or invalid we are not saying anything either about the premisses of the argument taken by themselves or about the conclusion taken by itself: in particular when we say that an argument is valid we are not implying that its conclusion is true; and when we say that an argument is invalid we are not implying that its conclusion is false.

3. Truth-functors, truth-values, truth-functional validity. In this section we attempt to give some idea of the nature and scope of truth-functional logic. Truth-functional logic is concerned with truth-functional validity. An argument is truth-functionally valid if and only if it exemplifies a valid truth-functional argument form. We must now explain what is meant when an argument form is described as truth-functional.

An argument form consists of two parts: a variable part and a constant part. The variable part is the set of labelled spaces which may be filled in in various ways; and we have seen that each space-labelling letter is known as *a variable*. The constant part is all the rest of the argument form; it may be thought of as divided into a number of distinct elements each of which is known as *a constant*. For example in the form:

(F) (1) p or q.
      (2) It is not the case that p.
      Therefore (3) q.

The variables are the letters *p*, *q*, each having two occurrences, and the constants are the expressions:

> ... or ... .
> It is not the case that ... .
> Therefore ... .

The constant *therefore* serves to mark the conclusion: it is not a part of any of the propositions of which an argument form is made up. In the kinds of argument form of which we have so far had instances and with which alone we shall here be concerned those constants which have not simply a conclusion or premiss marking function may be regarded as proposition formers. A proposition former is, so to speak, the framework, or part of the framework, of a proposition. By means of it and of other elements which we call associated components we are able to form or construct a proposition. For different formers different kinds of associated component are needed. For example, in the form A we may regard the expression: *Every ... is a ...* which is part of the first line as a proposition former; here for associated components we need general terms. On the other hand in the form D we may regard the expression: *Aristotle said that ...* as a proposition former but here as associated component we need not a general term but a proposition. We may say that *Aristotle said that ...* is a proposition former of propositional associated components.

10

Introduction

It can be seen that if a proposition former requires a certain kind of associated component, then if this proposition former is used in a form, part at least of the type rule will be that the appropriate variables are to be replaced by that kind of component. Other examples of proposition formers of propositional associated components are ... *before* ..., ... *or* ... and the two constants of this kind in the form F, namely: *It is not the case that* ... and *Both* ... *and* ... which we may shorten to ... *and* ... . These expressions have the property that by means of them we may form compound propositions out of more simple ones. Thus given the propositions:

(1) Hitler occupied the Rhineland,
(2) Mussolini invaded Abyssinia,

by means of *and* we may construct a new proposition:

(3) Hitler occupied the Rhineland and Mussolini invaded Abyssinia;

and by means of *or* we may construct the proposition:

(4) Mussolini invaded Abyssinia or Hitler occupied the Rhineland.

Again by means of the expression *It is not the case that* we may construct new propositions with, for example, (1) and (4) respectively as components, namely:

(5) It is not the case that Hitler occupied the Rhineland.

(6) It is not the case that Mussolini invaded Abyssinia or Hitler occupied the Rhineland.

A proposition which is itself compound may be a component of a compound proposition; thus the compound proposition (4) is a component of (6). Again the compound proposition (5) is a component of the following compound proposition:

B   11

(7) It is not the case that it is not the case that Hitler occupied the Rhineland.

We may say that *It is not the case that* is a monadic or one-place proposition former since it can be used to construct a compound proposition with a single proposition, either simple or compound, as component. *And* and *or* on the other hand are dyadic or two-place proposition formers, each of them being used to construct a compound proposition with two components.

And, *it is not the case that*, *Aristotle said that* and *before* are all proposition formers which use propositional components. However, the first two differ from the third and fourth in an important way. *And* and *it is not the case that* are, or at least may be used as, truth-functional proposition formers; *Aristotle said that* and *before*, on the other hand, are not truth-functional. We must now explain the property of truth-functionality which some proposition formers of propositional associated components have and others lack. A proposition former is truth-functional if the truth or falsity of a proposition formed by it with certain components depends in respect of those components solely on their truth or falsity. In other words, in order to know whether a proposition so formed is true or false we need never know any other fact about each of its components than whether that component is true or false. For example, in order to know whether the compound proposition (5) is true or false we need not know any other fact about its component, the proposition (1), than whether (1) is true or false; if (1) is true then (5) must be false and if (1) is false (5) must be true; the truth or falsity of the component is all that matters: we need not even know what it means, provided that we can know whether it is true or false without knowing its meaning. Again, in order to know whether the compound proposition (3) is true or false we need know no other facts about its components, the propositions (1) and (2), than facts concerning their truth or falsity. There are four possibilities only: (1) and (2) may both be true; (1) may be true and (2) false; (1) may be false and (2) true; and both may be false. All we need know is which of these cases obtains. If the first obtains then (3) is true; in any other case (3) must be false.

It is easy to show that not all proposition formers of propositional associated components have this property of truth-functionality. Consider two formers which we have mentioned earlier: *Aristotle said that* and *before*. By means of *before* we may construct the compound proposition:

(8) Hitler occupied the Rhineland before Mussolini invaded Abyssinia,

which has (1) and (2) as components. Again, given:

(9) The earth is flat,

we may use the expression *Aristotle said that* to construct the compound proposition:

(10) Aristotle said that the earth is flat.

These two proposition formers do not have the property of truth-functionality. In order to know whether the proposition (10) is true or false it is not enough that we know whether its component, the proposition (9), is true or false. We cannot say on just this information that (10) is true; for there are many true statements which Aristotle omitted to make, and also many false ones. Similarly, we cannot say that (10) is false, since of the statements made by Aristotle some were true and some were false. Again it is not the case that in order to decide about the truth or falsity of a statement such as (8) we need never know any other facts concerning its components than facts concerning their truth or falsity; for both components may be true and yet our knowing this will not enable us to know whether or not (8) is true: for obviously it is possible to

Introduction

know both that Hitler occupied the Rhineland and that Mussolini invaded Abyssinia and yet not to know whether or not the first-mentioned event preceded the second. It is clear then that some proposition formers of propositional associated components are not truth-functional.

The definition which we have given of the term *truth-functional* may be expressed more neatly if we introduce the term *truth-value*. This term is used in the following way : to ask what the truth-value is of a certain proposition is to ask whether that proposition is true or false; to say that a certain proposition has the truth-value, true, is to say that that proposition is true; and to say that it has the truth-value, false, is to say that it is false. Instead of the words *true* and *false* the numerals *1* and *0* are commonly used for the truth-values. Thus to say that a proposition has the truth-value 1 is to say that it is true and to say that it has the truth-value 0 is to say that it is false. We may now define truth-functionality thus: a proposition former is truth-functional if and only if, when a proposition is formed by it with certain components, the truth-value of that proposition depends, so far as its components are concerned, solely on their truth-values.

Our object in this section has been to give some account of the essential character of truth-functional logic. We said at the outset that truth-functional logic is concerned with truth-functional validity and that an argument is truth-functionally valid if, and only if, it exemplifies a valid truth-functional form. We are now able to complete this account by explaining what we mean in describing a form as truth-functional.

An argument form is truth-functional if and only if all the constants in its premises and conclusion are truth-functional proposition formers.

An example of a truth-functional argument form is the form F.

14

Introduction

We have said that an argument is truth-functionally valid if and only if it exemplifies a valid truth-functional form. It follows of course that if an argument does not exemplify a valid truth-functional form it is not truth-functionally valid. It may be convenient at times to speak of an argument which is not truth-functionally valid as being truth-functionally invalid. We must be careful, however, if we do so that we are not misled. If an argument is truth-functionally valid it is valid. But if an argument is truth-functionally invalid in the sense just stated, namely that of not being truth-functionally valid, it is not necessarily invalid; for though it is not truth-functionally valid it may yet be valid in some other way. For example the argument:

> Some footballers are cricketers.
> Therefore, some cricketers are footballers.

does not exemplify any valid truth-functional form. It is therefore not truth-functionally valid and hence, if we adopt the usage under discussion, is truth-functionally invalid. Despite this it is a valid argument exemplifying as it does the valid but not truth-functional form

> Some a are b.
> Therefore, some b are a.

In practice, however, when we have occasion to consider whether an argument is truth-functionally valid, very often, though not always, it is obvious that there can be no question of any kind of validity other than truth-functional validity: if in such a case we show that the argument is in fact truth-functionally invalid we naturally conclude that it is invalid absolutely.

The term *proposition former* has been used in this section for explanatory purposes. The symbols which we have called truth-functional proposition formers are, however, better known as *truth-functional constants* or as *truth-*

15

Introduction

functional operators or as *truth-functors*. We will hence-forward refer to them as truth-functors.

Later on we will use a number of special symbols which will be strictly defined and which will be employed exclusively as truth-functors. However, it may be helpful at this point to list some expressions of ordinary speech which have truth-functional uses. We have already mentioned *and* and *not*; to these may be added *if*, *or* and *neither ... nor ...* . We have given above a fairly careful statement of the field of truth-functional logic. If a less technical account should be wanted we might say, as a rough-and-ready guide but no more, that truth-functional logic studies arguments of which the force depends on some or all of the words *not*, *and*, *or*, *if* and their equivalents. Even in this apparently cautious statement, however, we are anticipating conclusions reached after discussion of a controversial question in chapter v, part ii.

The next chapter falls into two parts. The first part, which consists of sections 1 to 4, is devoted to an account of truth-functionality which will be rather more comprehensive than what has so far been attempted. The reader may, however, if he so wishes, proceed now to the second part of chapter ii. This is intended to be intelligible without the first part. It consists of sections 5 to 7 and contains a sufficient explanation of the symbols which will be used thereafter as truth-functors. Before proceeding to chapter ii, however, we introduce the term *propositional form*. A propositional form is an expression containing variables which becomes a proposition when substitutions are made for all variables. The premisses and the conclusions of the argument forms A—G are all propositional forms. A propositional form of course has propositions as its exemplifications; e.g., the first premiss of the argument form F is a propositional form and has as an exemplification the first premiss of the argument F_1.

Chapter Two

TRUTH-FUNCTIONS

Part I

1. The idea of a truth-function; notation for truth-functions. In the previous chapter we explained the meanings of two important expressions which contain the word *truth-functional*, namely *truth-functional constant* and *truth-functional argument form*. On the basis of these explanations and of the definitions which will be given in section 5 of a number of truth-functors or truth-functional constants it is possible to build a useful knowledge of a good part of our subject. However, for a wider understanding it is desirable that we should be on familiar terms not just with the adjective *truth-functional* but also with the noun *truth-function*.

It should be said at the outset that the word *truth-function* is sometimes used to stand simply for a proposition or a propositional form constructed by means of a truth-functor. This is not the sense which we have in mind here; we will refer to such expressions where necessary as *truth-functional propositions* or *truth-functional propositional forms*. We reserve the word *truth-function* for a different, though to some extent related, idea. In this chapter we will explain the idea of a truth-function in this distinct sense of the word and will mention some points of interest in relation to the truth-functions with which we shall be chiefly concerned.

A truth-function may be regarded as a set of pairs of elements such that (i) the first member of each pair is an

17

ordered set of truth-values and the second is a single truth-value and (ii) no two distinct pairs have the same first member. In each case the first member, i.e. the ordered set of truth-values, may be called an *argument* of the function and the second member may be called the *value of the function* for that argument. This is of course a different sense of the word *argument* from the one explained in the last chapter but there is not likely to be any danger of confusion. The whole class of arguments of a given function is known as the *domain* of that function and the class of values, i.e. of second members, is known as its *range*.

Each truth-function is said to be *n*-adic, where *n* is an integer greater than 0. An *n*-adic truth-function is a set of 2^n pairs of elements, the first member of each pair being an ordered set of *n* truth-values and the second member a single truth-value. For example, a 2-adic (dyadic) truth-function is a set of $2^2=4$ pairs of elements; one such function may be represented as follows:

	DOMAIN 1st member of pair (an ordered set of 2 truth-values)	RANGE 2nd member of pair (a single truth-value)
1st pair	$\{1, 1\}$	0
2nd pair	$\{1, 0\}$	1
3rd pair	$\{0, 1\}$	1
4th pair	$\{0, 0\}$	0

All distinct dyadic truth-functions represented in this way will have the same four items ($\{1, 1\}$, $\{1, 0\}$, $\{0, 1\}$, $\{0, 0\}$) in the domain column but will differ from one another in their range column. The order of the pairs themselves is indifferent since a function is a set of pairs not an ordered set of pairs; hence the following table represents exactly the same truth-function as that shown above:

18

Truth-Functions

DOMAIN	RANGE
$\{1, 0\}$	1
$\{1, 1\}$	0
$\{0, 0\}$	0
$\{0, 1\}$	1

It can be seen that there are as many dyadic truth-functions as there are ways of pairing the two possible second members, the truth-values 1 and 0, with the 2^2 possible first members, the ordered sets $\{1, 1\}$, $\{1, 0\}$, $\{0, 1\}$ and $\{0, 0\}$; that is to say there are $2^{2^2}(=16)$ 2-adic truth-functions. On the other hand there are only four 1-adic (monadic) truth-functions. Each 1-adic truth-function is a set of pairs of elements such that the first member of each pair is an ordered set of one truth-value and the second member is a truth-value; since the only possible first members are the $2^1=2$ sets each consisting of a single truth-value, i.e. the set consisting of the value 1 and the set consisting of the value 0, and the only possible second members are the two truth-values 1 and 0, it can be seen that there are $2^{2^1}=4$ ways of pairing possible second members with possible first members and hence four 1-adic truth-functions. In general there are 2^{2^n} n-adic truth-functions.

We must now adopt a notation for representing the different truth-functions. We will use the letter T with a subscript to represent a truth-function and there will be a distinct subscript for each distinct function. We will now explain how the subscripts are formed. Let us deal first with the 1-adic functions. Each of these is a set of two pairs. It is not, as we have seen, an ordered set of two pairs. Nevertheless, for the purpose of notation it is convenient to think of the pairs as occurring in a certain, arbitrarily chosen, order and we shall define this order by reference to the first members of the pairs. These first members are

19

the sets consisting of 1 and 0 respectively and we shall regard them as having the order: $\{1\}$, $\{0\}$. That is to say in each function the pair with first member $\{1\}$ will be regarded, for the purpose of notation, as preceding the pair with first member $\{0\}$. With this convention the different functions may be distinguished simply by reference to the second members of the pairs; and we shall use as subscript a row of two figures enclosed in brackets, the first figure representing the truth-value which is the second member of the *first* pair belonging to the function and the second figure representing the truth value which is the second member of the *second* pair belonging to the function. To put this otherwise, if x and y are truth-values, not necessarily distinct, $T_{(xy)}$ will be the function which consists of the pairs:

$$\text{and} \quad \begin{matrix} \{1\} & x \\ \{0\} & y. \end{matrix}$$

Thus the monadic truth-functions are:

$T_{(1\ 1)}$, consisting of the pairs $\{1\}$ 1 and $\{0\}$ 1,
$T_{(1\ 0)}$, consisting of the pairs $\{1\}$ 1 and $\{0\}$ 0,
$T_{(0\ 1)}$, consisting of the pairs $\{1\}$ 0 and $\{0\}$ 1,
and $T_{(0\ 0)}$, consisting of the pairs $\{1\}$ 0 and $\{0\}$ 0.

To represent dyadic truth-functions we again fix a standard order of first members. This will be $\{1, 1\}$, $\{1, 0\}$, $\{0, 1\}$, $\{0, 0\}$. We use as subscript to T for each function a row of four figures in brackets which represent in order from left to right the second members of the four pairs in which the function consists. Thus where w, x, y, z are truth-values the symbol denoting a dyadic truth-function will be of the form: $T_{(wxyz)}$. For particular cases of w, x, y, z this will represent the truth-function which consists of the pairs:

$$\{1, 1\} \quad w,$$
$$\{1, 0\} \quad x,$$
$$\{0, 1\} \quad y,$$
$$\{0, 0\} \quad z.$$

For example $T_{(1001)}$ is the function which consists of the pairs:

$$\{1, 1\} \quad 1$$
$$\{1, 0\} \quad 0$$
$$\{0, 1\} \quad 0$$
$$\{0, 0\} \quad 1$$

The following then is a complete list of the sixteen dyadic truth-functions:

$T_{(1111)}$, $T_{(1110)}$, $T_{(1101)}$, $T_{(1100)}$, $T_{(1011)}$, $T_{(1010)}$, $T_{(1001)}$, $T_{(1000)}$, $T_{(0000)}$, $T_{(0001)}$, $T_{(0010)}$, $T_{(0011)}$, $T_{(0100)}$, $T_{(0101)}$, $T_{(0110)}$, $T_{(0111)}$.

We now have a notation for all monadic and dyadic truth-functions. We shall not in this book have any occasion to refer to *n*-adic functions where $n > 2$.

Value of a function

In any function the truth-value which is paired with a particular element *e* of the domain is said to be the value of the function for *e* and is denoted by writing *e* immediately after the function sign. Thus $T_{(01)}\{1\}$ is the value for $\{1\}$ of the function $T_{(01)}$ and it is identical with 0 since 0 is the value which is paired with $\{1\}$ in this function. Again $T_{(1011)}\{0,1\}$ is the value for $\{0,1\}$ of the function $T_{(1011)}$ and is identical with 1 since 1 is the value which is paired with $\{0,1\}$ in this function.

2. Truth-functors and truth-functions. We now resume our discussion of truth-functional constants or truth-functors. A truth-functor is a symbol which is used to form a compound proposition out of one or more propositions. A truth-functor is *n*-adic, i.e. 1-adic, 2-adic and so on. A 1-adic truth-functor forms a new proposition out of a single proposition. A 2-adic truth-functor forms a new proposition out of two propositions or out of one proposition used twice; and so on. We will here use as our basic truth-functors the symbols which we have been using as subscripts in our notation for functions. Thus as monadic truth-functors we will use the symbols:

(11), *(10)*, *(01)* and *(00)*;

and as dyadic truth-functors we will use:

(1111), *(1110)*, *(1101)*, *(1100)*, *(1011)*, *(1010)*, *(1001)*, *(1000)*, *(0000)*, *(0001)*, *(0010)*, *(0011)*, *(0100)*, *(0101)*, *(0110)*, and *(0111)*.

A new proposition is formed when a monadic truth-functor is written in front of a single proposition, when a dyadic truth-functor is written in front of two propositions and so on. Thus if *p* and *q* are propositions then the following, for example, are also propositions: *(11)q*, *(01)p*, *(1110)pq*, *(0011)qp*, *(1001)(11)q(01)p*.

We must now define the truth-functors. We do this by explaining the meanings of the propositions which they form. With each truth-functor there is associated a single truth-function, namely the truth-function which, in the notation explained, has that truth-functor as subscript. For example, with *(00)* is associated the function $T_{(00)}$, with *(1110)* is associated the function $T_{(1110)}$. Each truth-functor may be defined by reference to its associated truth-function; the distinct definitions may all be covered by a number of comprehensive formulae, one for monadic truth-functors, one for dyadic truth-functors and so on.

Comprehensive definition for monadic truth-functors

If F_1 is a monadic truth-functor and *p* is a proposition and v_p is the truth-value of *p*, then the proposition: $F_1 p$

is equivalent to the assertion that the truth-value v_p of p is such that the value for $\{v_p\}$ of the function T_{F_1} is 1, i.e. that $T_{F_1}\{v_p\}=1$. For example, if F_1 is the functor *(01)* and p is the proposition: *Plato wrote the Republic* we see from the above definition that the meaning of the proposition:

(1)   (01) Plato wrote the Republic

is that the truth-value v_p of the proposition *Plato wrote the Republic* is such that $T_{(01)}\{v_p\}=1$. Since $T_{(01)}\{v_p\}=1$ if and only if $v_p=0$ the proposition (1) means that the proposition: *Plato wrote the Republic* has the truth-value **0**, i.e. is false. (1) then means that the proposition: *Plato wrote the Republic* is false. However, *Plato wrote the Republic* is in fact a true proposition; hence (1) is, as it happens, itself false.

Comprehensive definition for dyadic truth-functors

If F_2 is a dyadic truth-functor, p and q are propositions and v_p and v_q are the truth-values of p and q respectively then the proposition: $F_2 pq$ is equivalent to the assertion that v_p and v_q are such that the value for $\{v_p, v_q\}$ of the function T_{F_2} is 1, i.e. that $T_{F_2}\{v_p, v_q\}=1$. Consider, for example, the case in which F_2 is the functor *(1110)*, p is the proposition: *Plato wrote the Republic* and q is the proposition: *Aristotle wrote the Republic*; in accordance with our definition the proposition:

$$p \qquad\qquad\qquad\qquad\qquad\qquad q$$
(2)   (1110) Plato wrote the Republic Aristotle wrote the Republic

is equivalent to the assertion that the truth-values v_p and v_q of p and q are such that the value for $\{v_p, v_q\}$ of $T_{(1110)}$ is 1. Consider now the following tabular representation of $T_{(1110)}$.

23

$$T_{(1110)}$$

$\{1, 1\}$	1
$\{1, 0\}$	1
$\{0, 1\}$	1
$\{0, 0\}$	0

We see that $T_{(1110)}$ has the value 1 for $\{1, 1\}$, $\{1, 0\}$ and $\{0, 1\}$ but not for $\{0, 0\}$. The proposition (2) thus is equivalent to the assertion that one of the following three cases holds:

$$v_p=1, \ v_q=1$$
$$v_p=1, \ v_q=0$$
$$v_p=0, \ v_q=1 ;$$

in other words that at least one of the two propositions p and q is true. Since p is in fact true (2) is, as it happens, a true proposition.

Comprehensive definitions of n-adic truth-functors for cases of n greater than 2 can be formulated leading to a completely general statement which will apply for any value of n. A fully general account of truth-functions is, however, beyond the scope of the present text.

3. Individual monadic and dyadic truth-functions.

We will now make some remarks about the different monadic and dyadic truth-functors and their associated functions. We begin with the monadic functors: *(11)*, *(10)*, *(01)*, and *(00)*. The proposition *(11)p* is equivalent to the assertion that $T_{(11)}\{v_p\}=1$. But $T_{(11)}\{1\}=1$ and also $T_{(11)}\{0\}=1$; hence *(11)p* is true whether p is true or false; thus the assertion *(11)p* tells us nothing about the value of p. Similarly we learn nothing about the value of p from the proposition *(00)p*. *(00)p* is equivalent to the assertion that $T_{(00)}\{v_p\}=1$. But there is no value of p such that $T_{(00)}\{v_p\}=1$; hence *(00)p* is false whatever the value of p; it gives no information about p. The functions $T_{(11)}$ and $T_{(00)}$ may be called *constant* functions since the value

Truth-Functions

of each of them is constant for all elements in its domain. $T_{(11)}$ is a constant true function and $T_{(00)}$ is a constant false function. The former function is sometimes known as *Verum* and the latter as *Falsum*.

The proposition $(10)p$ is equivalent to the assertion that $T_{(10)}\{v_p\}=1$. Since this is the case if and only if $v_p=1$, i.e. p is true, it follows that $(10)p$ makes in effect the same assertion as p. $T_{(10)}$ may be known as an identity function. It is sometimes called by the name *Assertion*.

We now come to the functor (01). $(01)p$ is equivalent to the assertion that $T_{(01)}\{v_p\}=1$. This is the case if and only if $v_p=0$, i.e. p is false. Hence $(01)p$ is in effect equivalent to the assertion that the proposition p is false. It is clear, I think, that (01) is the most important and interesting of the monadic truth-functors; $(11)p$ and $(00)p$ say nothing; (10) is in a sense redundant since $(10)p$ says the same as p. But $(01)p$ is a definition assertion, distinct from p, and equivalent to the assertion that p is false. The function $T_{(01)}$ is known as *Negation*; the value for $\{v_p\}$ of $T_{(01)}$ may be said to be the negation of v_p, i.e. $T_{(01)}\{v_p\}$ is the negation of v_p: thus 0 is the negation of 1 and 1 is the negation of 0; the proposition $(01)p$ is also said to be the negation of the proposition p.

We come now to the dyadic truth-functors. For convenience of reference we set these out in eight columns, numbered 1 to 8, and two rows, lettered U and L.

	1	2	3	4	5	6	7	8
U	*(1111)*	*(1110)*	*(1101)*	*(1100)*	*(1011)*	*(1010)*	*(1001)*	*(1000)*
L	*(0000)*	*(0001)*	*(0010)*	*(0011)*	*(0100)*	*(0101)*	*(0110)*	*(0111)*

We may consider first the functors in the upper row. We see at once that 1, 4 and 6 are of little interest. $T_{(1111)}$ is the constant true dyadic function and $(1111)pq$ gives no information about p or q. $T_{(1100)}$ and $T_{(1010)}$ are identity functions of a kind. If v_1 and v_2 are truth-values it can be seen that $T_{(1100)}\{v_1, v_2\}$ is always identical with v_1 and

25

that $T_{(1010)}\{v_1, v_2\}$ is always identical with v_2; hence *(1100)pq* makes the same assertion as *p* and *(1010)pq* makes the same assertion as *q*. $T_{(1100)}$ and $T_{(1010)}$ might be known respectively as a first-element identity function and a second-element identity function. The counterparts in the lower row of these three functors are similarly comparatively unimportant. $T_{(0000)}$ is the constant false dyadic function. $T_{(0011)}$ and $T_{(0101)}$ are negations of a kind; *(0011)pq* makes the same assertion as *(01)p* and *(0101)pq* makes the same assertion as *(01)q*.

The remaining ten dyadic functors are those which will concern us most. We shall discuss first those in the upper row. If *p* and *q* are propositions and v_p and v_q are the truth values of *p* and *q* respectively, then the proposition *(1110)pq* is equivalent to the assertion that $T_{(1110)}\{v_p, v_q\} = 1$. The table for $T_{(1110)}$, i.e.

$$
\begin{array}{ll}
\{1, 1\} & 1 \\
\{1, 0\} & 1 \\
\{0, 1\} & 1 \\
\{0, 0\} & 0
\end{array}
$$

shows that $T_{(1110)}\{v_p, v_q\} = 1$ for three cases only of $\{v_p, v_q\}$ namely $\{1, 1\}$, $\{1, 0\}$ and $\{0, 1\}$. Thus *(1110)pq* is equivalent to the assertion that one of these three cases obtains, i.e. in effect that at least one of the propositions *p*, *q* is true. The functor *(1110)* corresponds to the word *or* or the expression *either ... or* in one common sense. The function $T_{(1110)}$ is known as *Disjunction* or *Alternation*; the proposition *(1110)pq* is known as the disjunction of *p* and *q* and we may also say that $T_{(1110)}\{v_q, v_p\}$ is the disjunction of v_p and v_q.

We consider next number 5 in the upper row. The proposition *(1011)pq* is equivalent to the assertion that $T_{(1011)}\{v_p, v_q\} = 1$. Since this is true for three cases only of $\{v_p, v_q\}$ namely $\{1, 1\}$, $\{0, 1\}$, $\{0, 0\}$, *(1011)pq* is equivalent to the assertion that one of these three cases

26

holds; in other words it excludes the possibility that p is true and q is false but leaves open all other possibilities. $T_{(1011)}$ is known as *Implication* or *Material Implication*. The proposition $(1011)pq$ is referred to as the material implication by p of q.

The functor (1101) can be seen to stand in a special relation to (1011). $(1011)pq$ excludes the possibility that p is true and q is false and leaves open all other possibilities; $(1101)pq$ on the other hand excludes the possibility that q is true and p is false and leaves open all other possibilities. $(1101)pq$ makes the same assertion as $(1011)qp$; it may be called the material implication by q of p or the converse of the implication by p of q. The function $T_{(1101)}$ may be called *Converse Implication* or *Converse Material Implication*.

$(1001)pq$ is equivalent to the assertion that $T_{(1001)}$ $\{v_p, v_q\}=1$. This is possible if but only if v_p and v_q are both 1 or both 0; hence $(1001)pq$ is in effect equivalent to the assertion that either p and q are both true or they are both false; the possibility that one is true and the other false is excluded. The function $T_{(1001)}$ is known as *Equivalence* or *Material Equivalence*. $(1001)pq$ may be called the equivalence of p and q or may be said to assert the equivalence of p and q.

$(1000)pq$ is equivalent to the assertion that $T_{(1000)}$ $\{v_p, v_q\}=1$. From the table for $T_{(1000)}$ it can be seen that this is possible solely when v_p and v_q are both 1, i.e. when p and q are both true. $(1000)pq$ has the same meaning as p *and* q or *Both* p *and* q in the normal sense of *and*. $(1000)pq$ is said to be the conjunction of p and q or to assert the conjunction of p and q, and the function $T_{(1000)}$ is called *Conjunction*.

The lower line functors may be dealt with more quickly. All lower line functors stand in the same relation to their counterparts on the upper line: where an upper line functor has *1*, the lower line one has *0* and where the upper line functor has *0*, the lower line functor has *1*. This means that if U is an upper line functor and L the corresponding lower line one the truth-value of *Lpq* is always the negation of the value of *Upq* and conversely. For example, the truth-value of *(0001)pq* is always the negation of the truth-value of *(1110)pq*; hence *(0001)pq* makes the same assertion as *(01)(1110)pq*. In general *Lpq* makes the same assertion as *(01)Upq*; and it is usual to regard the upper line functors as fundamental and to think of the lower line ones as defined by means of them. However, some of the lower line functors have a special interest of their own. For example, *(0001)pq* corresponds to *Neither p nor q* and *(0110)pq* corresponds to *Either p or q* in the so-called exclusive sense of *or*: it is true provided that *p* and *q* are neither both true nor both false.

We mentioned above that lower line functors may be defined by means of the functor *(01)* and the upper line functors. This, of course, is very far from exhausting the possibilities of interdefinability among truth-functors. Indeed it was shown by H. M. Sheffer that all possible truth-functors may be defined by means of the sole functor *(0111)* which he symbolized by means of a vertical stroke—*p | q* corresponding to *(0111)pq* in the notation which we have been using. It has also been shown that all truth-functors may be defined by means of the *neither-nor* functor *(0001)*. For a proof of this the reader is referred to W. V. Quine's *Mathematical Logic*, chapter i, sections 8 and 9.

We will return to the topic of interdefinability in the next chapter where certain important equivalences will be established.

4. Notations. Various different notations are in use for truth-functors. Some of these are shown in the following table in which the notation we have been using in this chapter is referred to as the '*0-1* notation'.

Truth-Functions

TRUTH-FUNC-TIONS	CORRESPONDING TRUTH-FUNCTORS (shown with propositions p, q as associated components)			
	0-1 notation	Hilbert-Ackermann notation	Peano-Russell notation	Łukasie-wicz notation
$T_{(01)}$	$(01)p$	$\bar{p}$	$\sim p$	Np
$T_{(1110)}$	$(1110)pq$	$p \vee q$	$p \vee q$	Apq
$T_{(1011)}$	$(1011)pq$	$p \rightarrow q$	$p \supset q$	Cpq
$T_{(1001)}$	$(1001)pq$	$p \sim q$	$p \equiv q$	Epq
$T_{(1000)}$	$(1000)pq$	$p \,\&\, q$	$p \cdot q$	Kpq

In the remainder of the book we shall be using exclusively the Peano-Russell notation and in the second part of this chapter an independent account will be given of the meaning and use of the symbols concerned. In the meantime one or two remarks of some general interest may be made.

In the first place we may notice one important difference between the *0-1* notation and the Łukasiewicz notation on the one hand and the Hilbert-Ackermann and Peano-Russell notations on the other hand. This difference is in respect of the dyadic functors: in the *0-1* and Łukasiewicz notations a dyadic functor is written in front of (i.e. to the left of) the associated components, whereas in the other two notations it is written between them. An effect of this is that in these two latter notations bracketing or some other device or convention is needed to prevent the ambiguity which would otherwise characterize many formulae. This point will be dealt with more fully in respect of the Peano-Russell notation in the next chapter.

In the second place it will be observed that in our table of notations we have given the corresponding symbols for

29

Truth-Functions

only four of the five main dyadic functors in the *0-1* notation which were shown on the upper line. The upper line functor omitted is *(1101)*. The other three notations do not in fact normally use any special symbol for this; for, since *(1101)pq* is equivalent to *(1011)qp*, it may be expressed in these other notations by $q \rightarrow p$, $q \supset p$ and Cqp respectively. However, the Łukasiewicz notation does in fact have a special letter for this purpose, namely *B*, and in the Hilbert-Ackermann and Peano-Russell notations the reversals of $\rightarrow$ and $\supset$ may appropriately be used if we wish. Thus we may extend our table by the following line for the sake of a certain completeness:

$$T_{(1101)} \qquad (1101)pq \qquad p \leftarrow q \qquad p \subset q \qquad Bpq$$

With regard to the lower line functors, formulae formed by these are usually represented by negations of formulae formed by the corresponding upper line functors; for example *(0001)pq* is expressed in the other notations as $\overline{pvq}$, $\sim[pvq]$ and *NApq* respectively. However, special devices may be used: in the Peano-Russell notation an appropriate one is to form a lower line functor by drawing a vertical line through the corresponding upper line functor. We should thus have the following equivalences:

$$
\begin{aligned}
p \curlyvee q &= (0001)pq \\
p \not\supset q &= (0100)pq \\
p \not\subset q &= (0010)pq \\
p \not\equiv q &= (0110)pq \\
p \mid q &= (0111)pq
\end{aligned}
$$

This has a rather pleasing side-effect: $T_{(0111)}$ is Sheffer's function; when we express the corresponding functor in the Peano-Russell notation in the way described we draw a vertical line through the dot in *p.q*; the line, as shown above, absorbs the dot, leaving us with the symbol well known as *Sheffer's stroke*.

30

Part II

5. Elementary truth-tables. In the previous part of this chapter an account was given of the idea of a truth-function and it was shown how any truth-functor may be defined by reference to an associated truth-function. It is more usual, however, to define truth-functors in a more direct way by means of what are called truth-tables. A truth-table is a table showing the truth-value of a proposition formed by means of one or more truth-functors for each possible set of values that can be assigned to the truth-functional components. Thus the simplest truth-table, one in which the proposition concerned contains only a single truth-functor, serves to define that functor in the sense that it displays the conditions under which a proposition formed by means of the functor is true and those under which it is false. A truth-table of this kind will be known as an *elementary truth-table*.

The principal truth-functors in the notation to be used in the remainder of this book are the following:

$$\sim, \text{v}, \supset, \equiv, \text{ and } .$$

$\sim$ is a monadic functor and the others are dyadic. Thus if p and q are propositions, $\sim p$ and $\sim q$ are propositions formed by the functor $\sim$, each of them having a single component, p in the one case and q in the other; $p\text{v}q$, $p \supset q$, $p \equiv q$ and $p.q$ are propositions formed by the dyadic functors, each of them having the two components, p and q. The truth functor $\sim$ may be defined by the following elementary truth-table which shows the truth-value of the proposition $\sim p$ for each possible value of the component p:

p	$\sim p$
1	0
0	1

The two possible truth values, 1 and 0, which the component p may have are shown in the left-hand column. The right-hand column gives the corresponding value of $\sim p$ in each case. Thus the table shows that $\sim p$ is a proposition which is false if p is true and true if p is false.

Each dyadic truth-functor is defined by a table which shows for each possible set of truth-values of two propositions p and q the truth-value of the proposition formed by the truth-functor in question with p and q as components. The truth-tables for the dyadic functors are now given.

TRUTH-TABLE FOR v

p	q	$p \vee q$
1	1	1
1	0	1
0	1	1
0	0	0

TRUTH-TABLE FOR .

p	q	$p \cdot q$
1	1	1
1	0	0
0	1	0
0	0	0

TRUTH-TABLE FOR $\supset$

p	q	$p \supset q$
1	1	1
1	0	0
0	1	1
0	0	1

TRUTH-TABLE FOR $\equiv$

p	q	$p \equiv q$
1	1	1
1	0	0
0	1	0
0	0	1

The information provided by the elementary truth-tables is sometimes given in alternative ways, for example by means of matrices or by means of truth equations. Truth equations will be set out and used in chapter iii.

6. Relations of truth-functors to expressions of

Truth-Functions

ordinary discourse. An expression of ordinary discourse is liable to vary to some extent from one type of context to another in respect of the way in which it is used and of the significance to be attached to it. The truth-functors:

$$\sim, \vee, \supset, \equiv, \text{ and } . ,$$

on the other hand are artificial symbols, exactly defined by means of the truth-tables and not admitting of variation in significance. In view of this, exact equivalences in all respects between truth-functors and expressions of ordinary language are perhaps not to be looked for. Nevertheless, the truth-functors we are dealing with do correspond in important ways to certain expressions of ordinary discourse in the sense that these expressions can be used truth-functionally though the truth-functional use may not be the only one. Subject to these general qualifications we may set out the following correspondences:

$\sim p$	corresponds to	*It is not the case that p*
		(which may be shortened to *not-p*),
$p \vee q$	corresponds to	*p or q*
		(in the inclusive sense of *or*[1]),
$p.q$	corresponds to	*p and q*,
$p \supset q$	corresponds to	*If p then q*,
$p \equiv q$	corresponds to	*If, and only if, p then q.*

Most people will probably accept readily enough the first three correspondences to the extent at any rate of allowing that *not*, *or* and *and* can be used in the senses indicated although *and* and *or* at least may also be used in senses which are not purely truth-functional. Thus $\sim p$ cor-

---

[1] The distinction between the exclusive and inclusive senses of *or* is that if *p* and *q* are both true *p or q* must be false if *or* is used in the exclusive sense but may (and, if purely truth-functional, must) be true if *or* is used in the inclusive sense.

responds to *It is not the case that p* which we may abbreviate
to *not-p*; for *It is not the case that p* is a statement which is
false if *p* is true and true if *p* is false. If the proposition
There will be rain tomorrow is true then *It is not the case that
there will be rain tomorrow* is false, while if the former
proposition is false the latter is true.

pvq corresponds to *p or q* or to *either p or q* in the
inclusive senses of *or* and *either ... or ...* . For example
the proposition:

(1) Either Smith is a member or he is the son of a
member

would normally be regarded as true if either or both of
the component propositions are true but as false if both of
them are false. *Either ... or* here corresponds exactly to v.
The proposition:

(2) Smith is a member v he is the son of a member

appears to have exactly the same sense as (1): the truth
table for v shows that (2) also is true if either or both of
the component propositions are true but false if both are
false.

p.q corresponds to *p and q* or *Both p and q*. For example
the proposition:

(3) Smith is a member and he is the son of a member

is true if both component propositions are true but is
false if either of them is false. As can be seen from the
truth-table for . exactly the same holds for the following
proposition formed by means of . with the same components:

(4) Smith is a member . he is the son of a member.

The correspondence indicated between ⊃ and *if* and
to some extent the associated one between ≡ and *if and
only if* are likely to be regarded with more suspicion. Some

would say that *if* does not have a truth-functional use at all, certainly not one corresponding to the truth-table for $\supset$. This view is important but cannot be dealt with briefly; we must leave consideration of it to a later chapter[1] where it will be discussed in the context of a broad consideration of the question of the applicability of truth-functional logic to arguments of ordinary discourse. In the meantime in order that, without question-begging, we may be able to illustrate our discussions by reference to ordinary discourse we will adopt a convention which may make things easier for those who feel unable to accept the equivalence of $\supset$ and *if*. When a proposition of ordinary discourse which we are using as an example is of the form *If p then q* it should be understood as being used in exactly the same sense as *not [p and not q]* and anyone who so wishes may mentally substitute the latter proposition for the former. If the truth-functional uses of *and* and *not* are accepted, we can see that this latter proposition has exactly the same meaning as $p \supset q$.

We may also adopt the convention that in our examples *not*, *and* and *or* will bear exactly the truth-table meanings of $\sim$, . and v respectively.

7. Bracketing. A truth-functionally compound proposition, that is one formed by a truth-functor with one or more associated components, may have for components propositions which are themselves truth-functionally compound. Thus for example if p, q, r and s are truth-functionally simple propositions we may use . to form the compound proposition $r.s$, and $\sim$ to form the compound proposition $\sim p$. We may then use . again or some other dyadic functor to form a proposition which has as its components the two compound propositions already created; or again which has one of these for one component and a simple proposition for the other. Now if we con-

---

[1] Chapter v, part ii.

struct propositions in the way described it will be found that some of them will be characterized by ambiguity unless we adopt some special device or convention to prevent this. For example the formula:

$$p.q\lor r,$$

in the absence of any special convention would be ambiguous: it would be impossible to say whether it should be understood as symbolizing the proposition formed by means of . from the propositions p and $q\lor r$ or as the proposition formed by $\lor$ from the propositions $p.q$ and r; these two propositions are of course different as is sufficiently shown by the fact that if p and q were both false but r were true the first-mentioned proposition would be false but the other one would be true. In this book our fundamental method will be to use brackets. The way in which these are used is likely to be in general obvious to anyone who is still reading this book. Thus we distinguish between the two compound propositions mentioned above in the following way. The first is:

$$p.[q\lor r]$$

and the second is:

$$[p.q]\lor r.$$

We shall adopt however three special conventions concerning the functors $\sim$, . and $\lor$ respectively. The component of the proposition formed by means of $\sim$ at any occurrence will be understood to be the shortest complete propositional expression which follows that occurrence of $\sim$. Thus:

$$\sim p\lor q$$

is to be understood as the proposition formed by $\lor$ from $\sim p$ and q rather than as the proposition formed by $\sim$ from $p\lor q$. This latter proposition is denoted by:

$$\sim[p\lor q].$$

If we did not have this convention we should have to express the first proposition thus:

$$[{\sim}p]{\vee}q.$$

The convention enables us to dispense with this pair of brackets.

Our convention regarding . is based on the fact that a proposition formed by . from p and $[q.r]$, i.e. the proposition $p.[q.r]$ has in all possible cases the same truth-value as the proposition formed by . from $[p.q]$ and r, i.e. as the proposition $[p.q].r$. Each proposition is true when p, q and r are all true and false in all other cases. Accordingly we adopt the convention that in either case we may omit the brackets and use the form:

$$p.q.r$$

which may be regarded as representing either proposition indifferently.

Our convention regarding $\vee$ is exactly parallel. $p{\vee}[q{\vee}r]$ has in all possible cases the same truth-value as $[p{\vee}q]{\vee}r$. Each is false if p, q and r are all false but true in all other possible cases. Accordingly brackets may be omitted and we may use:

$$p{\vee}q{\vee}r$$

to represent either proposition indifferently.

Other conventions which make possible a more extensive omission of brackets are frequently adopted but will not be used in this book. There are many different systems in use but normally each treatise or textbook specifies its own conventions clearly and there does not seem to be any need to describe any of them here. We ought perhaps, however, to mention at this point one convention which has not to do with the omission of brackets but is in quite common use. This is that the conjunction of two propositions is expressed not by a dot

37

but simply by juxtaposition: thus *pq* is used in place of our *p.q*.

SUGGESTED READING

Church, Alonzo: *Introduction to Mathematical Logic* (1956), introduction.

Menger, K.: *Calculus* (1955), chapter iv.

Quine, W. V.: *Mathematical Logic*, chapter i.

Prior, A. N.: *Formal Logic* (1955), chapter i.

Basson, A. H., and O'Connor, D. J.: *Introduction to Symbolic Logic*.

Bochenski, I. M.: *A Précis of Mathematical Logic*.

ukasiewicz, J.: *Aristotle's Syllogistic*, chapter iv.

Nidditch, P. H.: *Elementary Logic of Science and Mathematics*, section 4.09.

Chapter Three

THE TRUTH-TABLE METHOD

1. Introduction. In this and the next two chapters we will consider some methods which may be used to test whether an argument or argument form is truth-functionally valid or invalid. Three methods will be dealt with:

> the truth-table method,
> the deductive method
> and the normal form method.

It will be shown that essentially the same methods may be used for another purpose also: namely to test whether or not a proposition or a propositional form is or is not logically true.[1] Our own interest, however, will be primarily in validity and invalidity. The present chapter will explain the truth-table method.

2. The use of truth-tables to discover the truth-value of a proposition. We will begin by defining some expressions which will be useful to us from time to time. A proposition or propositional form Q may be said to be *truth-functional with respect to a set S of component propositions or propositional forms P_1, P_2, ..., P_n* if and only if either (i) the set S contains a single member only which is identical with Q, or (ii) Q can be built up by a step-by-step process in which one starts with the members of S and at each step constructs a new propositional element (proposition or propositional form) by means of a truth-

---

[1] The deductive method, as described in this book, is used only to establish validity or logical truth, not invalidity or logical falsity.

functor and one or more propositional elements already available, it being permissible to use any one element any number of times. For example, the propositional form (A) $[p \supset q] \supset [[q \supset r] \supset [p \supset r]]$ is truth-functional (as indeed is every proposition or propositional form, in accordance with (i) of the above definition) with respect to itself; it is also truth-functional with respect to components $p \supset q$, $[[q \supset r] \supset [p \supset r]]$; thirdly it is truth-functional with respect to components $p \supset q$, $q \supset r$, $p \supset r$; fourthly it is truth-functional with respect to components p, q, r; and it is truth-functional with respect to certain other sets of components also.

A proposition or propositional form is *truth-functionally compound* if it is formed by means of a truth-functor with the appropriate number of associated components; otherwise it is *truth-functionally atomic*. For example, the propositions: (a) *Smith was condemned* and (b) *Robinson said that* $\sim$ *[Smith was guilty]* are truth-functionally atomic but

(c) Smith was condemned . Robinson said that $\sim$ [Smith was guilty]

is truth-functionally compound. In the case of propositional forms propositional variables are to be regarded as truth-functionally atomic.

We will now explain how truth-tables or equivalent definitions of the truth-functors may be used to determine the truth-value of a proposition which is truth-functional with respect to a certain set of components in a case where the truth-value of each of these components is known. The method has a number of possible variants. In the one which we shall describe it is possible to refer directly to the elementary truth-tables set out in section 5 of chapter ii, but it is more convenient perhaps to use certain truth-equations which can be constructed by reference to the truth-tables and give equivalent information. The truth-equations for the principal truth-functors are these:

The Truth-Table Method

$\sim 1 = 0$	$1v1 = 1$	$1.1 = 1$	$1 \supset 1 = 1$	$1 \equiv 1 = 1$
	$1v0 = 1$	$1.0 = 0$	$1 \supset 0 = 0$	$1 \equiv 0 = 0$
$\sim 0 = 1$	$0v1 = 1$	$0.1 = 0$	$0 \supset 1 = 1$	$0 \equiv 1 = 0$
	$0v0 = 0$	$0.0 = 0$	$0 \supset 0 = 1$	$0 \equiv 0 = 1$

There is a set of equations for each truth-functor, a set of two equations for $\sim$ and a set of four equations for each of the dyadic functors. The equations for $\sim$ show the truth-value of a proposition formed by $\sim$ for each of the two possible truth-values which the sole component for $\sim$ might have. The equations for each dyadic functor f show the truth-value of a proposition formed by f for each possible set of values which the two components for f might have. I think that this is done in a sufficiently perspicuous way. We may proceed therefore to describe our procedure for arriving at the truth-value of a proposition. In our explanation we will make use of some examples and for the purpose of these examples we will use certain letters to stand for certain propositions as follows: $H = Plato$ was a $Greek$, $J = Sophocles$ $wrote$ the Oedipus Tyrannus, $K = The$ Aeneid is $longer$ $than$ the Iliad, $L = Lucretius$ $lived$ $during$ the $reign$ of $Nero$. Both H and J are true propositions; whereas K and L are false.

Suppose now that we wish to find the truth-value of the compound proposition:

(1) $H.[Kv\sim L]$.

(1) is truth-functional in respect of the components H, K and L the truth-values of all of which are known. We first of all write down the proposition at the top of a table and to the left we write the components; under each component we put its truth-value; thus we obtain the table:

H	K	L	$H.[Kv\sim L]$
1	0	0	

The Truth-Table Method

We are now ready to work out the truth-value of (1). In the bottom right-hand corner of the table we copy down the formula from the upper line but replace the propositional letters by the appropriate truth-value figures as shown in the left-hand column. This gives us:

H	K	L	$H.[Kv{\sim}L]$
1	0	0	$1.[0v \sim 0]$

We now follow a step-by-step reductive procedure. At each step for one or more parts of the formula last arrived at we substitute its equivalent or their equivalents as shown in the truth-equations. Thus we start with the formula $1.[0v{\sim}0]$. No equation has this whole formula on its left-hand side: accordingly we cannot find a single value for it in one step. The parts of the whole formula are 1 and $0v{\sim}0$. There is no equation with $0v{\sim}0$ on its left-hand side. However, part of this latter formula, namely ${\sim}0$, is the left-hand side of an equation; accordingly our first step will be to write down the original formula with the right-hand side of that equation substituted for ${\sim}0$ and to show the formula so formed as equated to the original one. Thus we have:

$$1.[0v{\sim}0]=1.[0v1].$$

This completes the first step. Now the part $0v1$ of our new formula is the left-hand side of a truth equation; replacing it by the right-hand side we obtain 1.1. We equate this formula to the one last obtained and now have:

$$1.[0v \sim 0]=1.[0v1]=1.1.$$

This completes the second step. The truth-equation $1.1=1$ now enables us to complete the procedure; we end with:

$$1.[0v \sim 0]=1.[0v1]=1.1=1.$$

The truth-value of the original proposition (1) is now

42

given at the extreme right-hand side. The justification of this procedure is not difficult. It might be put thus.

In the first step we show that the truth-value of the proposition (1) $H.[Kv{\sim}L]$ is identical with the truth-value of a proposition $H.[Kvp_1]$ where p_1 is a proposition with truth-value 1. In the second step we show that the truth-value of $H.[Kvp_1]$ is identical with the truth-value of $H.p_2$ where p_2 has the truth-value 1. In the third step we show that the truth-value of $H.p_2$ is identical with the truth-value of a proposition p_3 where p_3 has the value 1. Finally we conclude that the truth-value of (1) is identical with the truth-value of p_3 and so is 1.

We now show the table which results when we apply this method to another proposition:

(2) $\sim[H{\supset}[[{\sim}KvJ]\equiv{\sim}H]]$.

H	J	K	$\sim[H{\supset}[[{\sim}KvJ]\equiv{\sim}H]]$
1	1	0	$\sim[1{\supset}[[{\sim}0v1]\equiv{\sim}1]]=\sim[1{\supset}[[1v1]\equiv0]]$ $=\sim[1{\supset}[1\equiv0]]=\sim[1{\supset}0]=\sim0=1.$

It will be noted that in this example it is possible to make two replacements at the first step: 1 for ~0 and 0 for ~1. Again, as of course always in this method, the truth-value of the proposition appears at the extreme right-hand side.

We are now in a position to explain the truth-table method of determining validity or invalidity. We shall explain in fact two variants of the method, to be called respectively the Direct Truth-table Method and the Indirect Truth-table Method.

3. Determining the validity or invalidity of an argument form by truth-tables: the direct method.

A truth-functional argument form is valid if and only if no possible exemplification has all its premisses true but its conclusion false; otherwise it is invalid. An exemplification of an argument form is the same as the form except that each distinct variable in the form is, in the exemplification, replaced consistently by the same proposition. For any given argument form there is of course no limit to the number of exemplifications. However, each exemplification is constructed by replacing each variable in the form by a proposition and every proposition is true or false. Hence it can be seen that the number of ways in which an exemplification of a form can be constructed is limited in relation to the possible truth-values of the replacements for variables. Thus if a form F contains a single variable (which may, of course, occur several times) there are two ways in which an exemplification may be constructed: according as the single variable is replaced by a true proposition or by a false one. If F contains two distinct variables α and β (with any number of occurrences) there are $2^2=4$ different ways in which an exemplification of F may be constructed: (i) α and β may both be replaced by true propositions; (ii) α may be replaced by a true proposition and β by a false one; (iii) α may be replaced by a false proposition and β by a true one; and (iv) α and β may both be replaced by false propositions. Each such way of constructing an exemplification of F corresponds to a way of distributing truth-values over the variables (or atomic components) of F. It can be seen that in general if there are n distinct variables of a form F there are 2^n such distributions of truth-values and so 2^n ways in which an exemplification of F may be constructed. Now, since each way W of constructing an exemplification of F corresponds to a distribution of truth-values over the variables of F we can always discover the truth-values of premisses and conclusion of an exemplification constructed in the way W. If then we work out these truth-values for each possible W we shall be able to discover whether or not F is a valid form. For if there is no way of constructing an exemplification of F in which all the premisses are true and the conclusion false then F is a valid argument form; but otherwise it is invalid.

We may illustrate this by some examples. Consider first the argument form:

(A) $p \supset \sim q$; therefore $q \supset \sim p$.

This form contains two distinct variables p and q. There are thus $2^2=4$ different ways in which an exemplification of A may be constructed, according as p is replaced by a true proposition or by a false one and as q is replaced by a true proposition or a false one. We may exhibit these four possible ways of constructing an exemplification of A in the following manner. We begin by constructing the table shown below. In the middle column the premiss of A is written and in the right-hand column its conclusion. At the head of the left-hand column are shown the two variables, p and q, which occur in A and underneath these the four different distributions of truth-values which are possible when p and q are replaced by propositions.

		PREMISS	CONCLUSION
p	q	$p \supset \sim q$	$q \supset \sim p$
1	1		
1	0		
0	1		
0	0		

Each line of the body of this table is now used to work out the truth-values which any exemplification of premiss and conclusion must have that is constructed in accordance with the distribution of truth-values shown in the left-hand column for that line. Thus on the top line we work out the truth-values of premiss and conclusion respectively for any exemplification which is constructed by replacing p by a true proposition and q by a true proposition; on the second line we do the same thing for any exemplification which is constructed by replacing p by a true proposition and q by a false one; and so on. The working out of truth-values may be done in the way explained in the previous section. We show the completed table.

		PREMISS	CONCLUSION
p	q	$p \supset \sim q$	$q \supset \sim p$
1	1	$1 \supset \sim 1 = 1 \supset 0 = 0$	$1 \supset \sim 1 = 1 \supset 0 = 0$
1	0	$1 \supset \sim 0 = 1 \supset 1 = 1$	$0 \supset \sim 1 = 0 \supset 0 = 1$
0	1	$0 \supset \sim 1 = 0 \supset 0 = 1$	$1 \supset \sim 0 = 1 \supset 1 = 1$
0	0	$0 \supset \sim 0 = 0 \supset 1 = 1$	$0 \supset \sim 0 = 0 \supset 1 = 1$

We see from this completed table that there is no way of constructing an exemplification of A in which the premiss is true and the conclusion false. If the first distribution is used we obtain a false premiss; if any other distribution is used we obtain a true premiss but also a true conclusion. It is clear then that no exemplification of the argument form A has its premiss true but its conclusion false. In accordance with our definition of validity A is therefore a valid argument form.

As a second example we may consider the argument form:

(B) $p \vee \sim q$, $q \supset r$; therefore $p \supset r$.

To test B for validity we use the same framework as before. However, there are two differences between A and B to be noted at the outset. In the first place B has two premisses whereas A had only one. To test for validity by the present method an argument form which has several premisses we must first of all use the functor . to conjoin all the premisses into a single propositional form.

46

The Truth-Table Method

To test B we put in our table immediately under the heading PREMISS the propositional form which is the conjunction of the premisses of B, i.e. the form:

$$[p \vee \sim q].[q \supset r].$$

In effect what we test immediately is the validity not of B but of the argument form

(B') $[p \vee \sim q].[q \supset r]$; therefore $p \supset r$,

which has a single premiss. The truth-table for . however should make it quite obvious that B is a valid argument form if and only if B' is also. B differs from A also in the number of distinct variables which it contains, three as compared with two in A. There are consequently $2^3 = 8$ possible distributions of truth-values and so there are eight (and only eight) possible ways in which an exemplification of B may be constructed. The eight possible distributions may be quickly written down as follows. Under r in the extreme left-hand column of our table we write 1's and 0's alternately; under q we write 1's and 0's in twos alternately, and under p, 1's and 0's in fours alternately. We now show the completed table for B'.

$p \; q \; r$	PREMISS $[p \vee \sim q].[q \supset r]$	CONCLU-SION $p \supset r$
1 1 1	$[1 \vee \sim 1].[1 \supset 1] = [1 \vee 0].1 = 1.1 = 1$	$1 \supset 1 = 1$
1 1 0	$[1 \vee \sim 1].[1 \supset 0] = [1 \vee 0].0 = 1.0 = 0$	$1 \supset 0 = 0$
1 0 1	$[1 \vee \sim 0].[0 \supset 1] = [1 \vee 1].1 = 1.1 = 1$	$1 \supset 1 = 1$
1 0 0	$[1 \vee \sim 0].[0 \supset 0] = [1 \vee 1].1 = 1.1 = 1$	$1 \supset 0 = 0$
0 1 1	$[0 \vee \sim 1].[1 \supset 1] = [0 \vee 0].1 = 0.1 = 0$	$0 \supset 1 = 1$
0 1 0	$[0 \vee \sim 1].[1 \supset 0] = [0 \vee 0].0 = 0.0 = 0$	$0 \supset 0 = 1$
0 0 1	$[0 \vee \sim 0].[0 \supset 1] = [0 \vee 1].1 = 1.1 = 1$	$0 \supset 1 = 1$
0 0 0	$[0 \vee \sim 0].[0 \supset 0] = [0 \vee 1].1 = 1.1 = 1$	$0 \supset 0 = 1$

The Truth-Table Method

It can be seen from the completed table that in the case of one distribution, namely the fourth shown, the premiss has the value 1 and the conclusion the value 0. It follows that in any exemplification of the argument form B' in which p is true but q and r are both false the premiss will be true and the conclusion false; since there are of course many such exemplifications the argument form B' is invalid. Consequently B also is invalid.

4. Shortened procedures. The reader should of course, practise the method we have explained. If he does so he will probably devise for himself time-saving modifications. We will explain here one procedure by which the work involved in this direct method may often be greatly reduced. Let us assume that we are dealing with an example of an argument form in which the conclusion is shorter than the conjunction of the premisses. We prepare our framework; we fill in the headings: the variables, the conjunction of the premisses, the conclusion. Under the variables we set out the possible distributions of truth-values. At this point we have nothing in the main part of the table, under either the premiss formula or the conclusion formula. We proceed now by stages. In Stage I we deal with the first line, i.e. with the first distribution of truth-values. We begin with the conclusion column and work out the truth-value of the conclusion. If this value is 1 we ignore the premiss in the first line and proceed to Stage II. If the value is 0 on the other hand we work out the truth-value of the premiss also in the first line; if this turns out to be 1 our work is over; for we have found that an exemplification can be constructed in which premiss is true and conclusion false and the argument form is therefore invalid. If, on the other hand, this value is 0 we proceed to Stage II. We repeat the same programme at Stage II in respect of the second line and go on in a similar way until either we show the argument form to be invalid or have completed the programme for every distribution. In the latter case, of course (i.e. if we have completed the programme for every distribution) if the argument form has not been shown to be invalid we know that it must be valid. We now show what difference this procedure would make to our work in the case of the argument form B'; the table we show below, which has been dealt with completely in accordance with this shortened procedure, should be compared with that on a previous page in which the unshortened procedure has been used.

$p\ q\ r$	PREMISS $[p \mathrm{v} \sim q].[q \supset r]$	CONCLU- SION $p \supset r$
1 1 1		$1 \supset 1 = 1$
1 1 0	$[1\mathrm{v} \sim 1].[1 \supset 0] = [1\mathrm{v}0].0 = 1.0 = 0$	$1 \supset 0 = 0$
1 0 1		$1 \supset 1 = 1$
1 0 0	$[1\mathrm{v} \sim 0].[0 \supset 0] = [1\mathrm{v}1].1 = 1.1 = 1$	$1 \supset 0 = 0$
0 1 1		
0 1 0		
0 0 1		
0 0 0		

For the first distribution the conclusion turns out to have the truth-value 1; so we proceed to the second distribution. Here the conclusion has the value 0 and we work out the value of the premiss which is also 0; and we go on to the third distribution. The conclusion has the value 1 and we proceed to the fourth case. Here the conclusion has the value 0. We work out the value of the premiss. Since this turns out to be 1 we know that there is a possible exemplification of B in which premisses are true and conclusion false and that B is therefore an invalid argument form.

A variant of this shortened method which is likely to be preferable if the premiss is shorter than the conclusion involves beginning at each stage with the premiss; the value of the conclusion needs to be discovered in each line if the value of the premiss is 1 but not otherwise.

5. An alternative idiom. We may now describe a way of speaking about exemplifications of argument or propositional forms which is in common use and which is for some purposes at least more convenient than that which we have been following in the present chapter. Instead of asking for example what properties an exemplification e of a form F would have if it were constructed (or constructible) by replacing a variable a by a true proposition we may ask simply what properties F would have if a were true. For example, on page 45 we wrote with reference to a form A:

There are thus $2^2=4$ different ways in which an exemplification of A may be constructed, according as p is replaced by a true proposition or by a false one and as q is replaced by a true proposition or a false one.

In the alternative idiom this sentence might be replaced by:

There are thus $2^2=4$ possibilities about A according as p is true or false and according as q is true or false.

The relationship between these two ways of speaking is interesting but cannot be gone into here. From our point of view we may regard the second, which we use to some extent, as a substitute for the first. It would not be right, however, to think of this second idiom as being in general necessarily in some way loose or incorrect. It is rather that in it we are speaking *by means of*, or *through* or *in* forms rather than directly *about* forms.

The Truth-Table Method

6. The indirect method. The direct method of determining the validity or invalidity of a truth-functional argument form involves, as we have seen, setting out each possible distribution of truth-values which an exemplification of the form might have and then discovering for each case the truth-values of the premiss and conclusion respectively. If we find that there is no case in which premiss is true and conclusion false then we know that the argument form is valid; if there is at least one case in which premiss is true and conclusion false it is invalid. Now consider for a given argument form F the propositional form which is constructed by means of the functor $\supset$ with premiss and conclusion of the argument form as components, the premiss being antecedent and the conclusion consequent. With P_F for the premiss of the argument form F and C_F for the conclusion we may represent this propositional form thus:

$$P_F \supset C_F.$$

The argument form F is valid if and only if there is no case in which the premiss of F is true and the conclusion of F is false; in other words if and only if there is no case in which P_F has the value 1 and C_F has the value 0. But we know from the truth-table for $\supset$ that there is no case in which P_F has the value 1 and C_F the value 0 if and only if there is no case in which $P_F \supset C_F$ has the value 0. Hence the argument form F is valid if and only if there is no possible case in which an exemplification of the propositional form $P_F \supset C_F$ has the value 0, i.e. is false. This relationship just stated between the validity of an argument form F and the possible truth-values of exemplifications of the propositional form $P_F \supset C_F$ is the basis of the indirect truth-table method of determining validity or invalidity. We construct for F the propositional form $P_F \supset C_F$; we set out all possible cases of distributions of truth-values in exemplifications of $P_F \supset C_F$ and work out

51

the truth-value of the exemplification for each possible case. If in all possible cases the truth-value is 1 the argument form F is valid; but if in one or more cases the truth-value is 0 then F is invalid.

We now describe the indirect method in more detail. Given an argument form F to test F for truth-functional validity by the indirect method we proceed as follows. First, if F has more than one premiss we combine all the premisses into one propositional form by conjunction; we call this newly constructed propositional form *the premiss of F*. If F has a single propositional form as premiss to start with, this of course will be *the premiss of F*; we use the expression P_F to denote the premiss of F in either case and we use the expression C_F to denote the conclusion. We now construct the propositional form which is the implication by P_F of C_F, ie.:

$$P_F \supset C_F.$$

We call this the corresponding implication for F or the implication corresponding to F and use CI_F as an abbreviation. Thus we may say:

$$CI_F = P_F \supset C_F.$$

We now find the truth-value of CI_F for each possible distribution of truth-values. We do this by a truth-table method which at this stage needs little explanation; it will emerge clearly in the examples which follow. If CI_F turns out to have the truth-value 1 in every case, i.e. for every possible distribution of truth-values, we conclude, for reasons explained above, that F is a valid argument form. If on the other hand in one or more cases CI_F has the truth-value 0, we conclude that F is an invalid argument form.

We now give some examples. Consider the argument form:

The Truth-Table Method

(D) $p \supset q$; $\sim q$; therefore $\sim p$.

P_D is $[p \supset q] . \sim q$. C_D is $\sim p$. So CI_D is:

$$[[p \supset q] . \sim q] \supset \sim p.$$

The truth-values of CI_D in all possible cases are worked out in the following table:

p	q	CORRESPONDING IMPLICATION for D (CI_D) $[[p \supset q] . \sim q] \supset \sim p$
1	1	$[[1 \supset 1] . \sim 1] \supset \sim 1 = [1.0] \supset 0 = 0 \supset 0 = 1$
1	0	$[[1 \supset 0] . \sim 0] \supset \sim 1 = [0.1] \supset 0 = 0 \supset 0 = 1$
0	1	$[[0 \supset 1] . \sim 1] \supset \sim 0 = [1.0] \supset 1 = 0 \supset 1 = 1$
0	0	$[[0 \supset 0] . \sim 0] \supset \sim 0 = [1.1] \supset 1 = 1 \supset 1 = 1$

CI_D is shown to have the truth-value 1 in all possible cases. Hence the argument form D is valid.

On the other hand the argument form:

(E) $p \supset q$, $\sim p$; therefore $\sim q$

is invalid. We show this by means of the following table in which we see that in one possible case CI_E has the value 0.

E is the form of the well-known fallacy, *Denying the antecedent.* Other notoriously fallacious forms are: *Affirming the consequent:* $p \supset q$, q, therefore p; *Fallacious hypothetical syllogism:* $p \supset q$, $r \supset q$, therefore $p \supset r$; or $q \supset p$, $q \supset r$, therefore $p \supset r$.

p	q	CI_E $[[p \supset q] . \sim p] \supset \sim q$
1	1	$[[1 \supset 1] . \sim 1] \supset \sim 1 = [1.0] \supset 0 = 0 \supset 0 = 1$
1	0	$[[1 \supset 0] . \sim 1] \supset \sim 0 = [0.0] \supset 1 = 0 \supset 1 = 1$
0	1	$[[0 \supset 1] . \sim 0] \supset \sim 1 = [1.1] \supset 0 = 1 \supset 0 = 0$
0	0	$[[0 \supset 0] . \sim 0] \supset \sim 0 = [1.1] \supset 1 = 1 \supset 1 = 1$

The Truth-Table Method

Again we see below that the argument form:

(G) $p \supset q$, $q \supset r$; therefore $p \supset r$,

is valid since CI_G is true in all cases.

			CI_G
p	q	r	$[[p \supset q] . [q \supset r]] \supset [p \supset r]$
1	1	1	$[[1 \supset 1] . [1 \supset 1]] \supset [1 \supset 1] = [1 . 1] \supset 1 = 1 \supset 1 = 1$
1	1	0	$[[1 \supset 1] . [1 \supset 0]] \supset [1 \supset 0] = [1 . 0] \supset 0 = 0 \supset 0 = 1$
1	0	1	$[[1 \supset 0] . [0 \supset 1]] \supset [1 \supset 1] = [0 . 1] \supset 1 = 0 \supset 1 = 1$
1	0	0	$[[1 \supset 0] . [0 \supset 0]] \supset [1 \supset 0] = [0 . 1] \supset 0 = 0 \supset 0 = 1$
0	1	1	$[[0 \supset 1] . [1 \supset 1]] \supset [0 \supset 1] = [1 . 1] \supset 1 = 1 \supset 1 = 1$
0	1	0	$[[0 \supset 1] . [1 \supset 0]] \supset [0 \supset 0] = [1 . 0] \supset 1 = 0 \supset 1 = 1$
0	0	1	$[[0 \supset 0] . [0 \supset 1]] \supset [0 \supset 1] = [1 . 1] \supset 1 = 1 \supset 1 = 1$
0	0	0	$[[0 \supset 0] . [0 \supset 0]] \supset [0 \supset 0] = [1 . 1] \supset 1 = 1 \supset 1 = 1$

7. Tautologies. In the indirect method which has just been described for testing whether an argument form F is valid or invalid we examine a certain truth-functional propositional form, CI_F, to see whether or not it has the property of being true for all possible distributions of truth-values over its variables. If it has this property it is said to be a *tautology* or to be *tautologous*. We may accordingly describe the essential part of the indirect method by saying that it consists in examining CI_F to see whether or not it is tautologous. However, the use of the term *tautology* is not restricted to this particular context. Any propositional form which has the property mentioned, whether or not it is an implication, is a tautology. We give the following definition.

A propositional form is a tautology, or a tautologous propositional form, if and only if its exemplifications are true for every possible distribution of truth-values over variables.

The Truth-Table Method

Not only propositional forms but also propositions may be described as tautologous.

> A proposition is a tautology, or a tautologous proposition, if and only if it is an exemplification of a tautologous propositional form.

It will be found that in the more advanced development of logic interest tends to be focused less on validity and more on what are variously called logically true propositional forms, logical laws or logical truths. A tautology is one kind of logical truth; we may say that it is a truth-functional logical truth. We give now a few simple examples of the use of truth-tables merely to test whether or not a propositional form is tautologous without any reference to a corresponding argument form.

(1) (H) $p \vee \sim p$.

p	H
	$p \vee \sim p$
1	$1 \vee \sim 1 = 1 \vee 0 = 1$
0	$0 \vee \sim 0 = 0 \vee 1 = 1$

(H) is tautologous.

(2) (I) $\sim [p . \sim p]$.

p	I
	$\sim [p . \sim p]$
1	$\sim [1 . \sim 1] = \sim [1 . 0] = \sim 0 = 1$
0	$\sim [0 . \sim 0] = \sim [0 . 1] = \sim 0 = 1$

(I) is tautologous.

(3) (L) $[p \supset q] \equiv]q \supset p]$.

		L
p	q	$[p \supset q] \equiv [q \supset p]$
1	1	$[1 \supset 1] \equiv [1 \supset 1] = 1 \equiv 1 = 1$
1	0	$[1 \supset 0] \equiv [0 \supset 1] = 0 \equiv 1 = 0$
0	1	$[0 \supset 1] \equiv [1 \supset 0] = 1 \equiv 0 = 0$
0	0	$[0 \supset 0] \equiv [0 \supset 0] = 1 \equiv 1 = 1$

(L) is not tautologous.

(4) (M) $[p \supset q] \equiv [\sim q \supset \sim p]$.

		M
p	q	$[p \supset q] \equiv [\sim q \supset \sim p]$
1	1	$[1 \supset 1] \equiv [\sim 1 \supset \sim 1] = 1 \equiv [0 \supset 0] = 1 \equiv 1 = 1$
1	0	$[1 \supset 0] \equiv [\sim 0 \supset \sim 1] = 0 \equiv [1 \supset 0] = 0 \equiv 0 = 1$
0	1	$[0 \supset 1] \equiv [\sim 1 \supset \sim 0] = 1 \equiv [0 \supset 1] = 1 \equiv 1 = 1$
0	0	$[0 \supset 0] \equiv [\sim 0 \supset \sim 0] = 1 \equiv [1 \supset 1] = 1 \equiv 1 = 1$

(M) is a tautology.

8. Logical equivalence; interdefinability of functors. This last example introduces us to an interesting and important class of tautologies. This is the class of tautologies in which the main functor is $\equiv$. Their importance is connected with the concept of *logical equivalence*, which may be defined by reference to them. Two propositional forms may be logically equivalent and so may two propositions. We may avoid the necessity of defining logical equivalence twice by introducing at this point the term *formula*.

An expression is a formula if and only if it is either a proposition or a propositional form.

The Truth-Table Method

Logical equivalence is now defined:

Two formulae F and G are logically equivalent if and only if the formula:

is a tautology. $\qquad$ $F \equiv G$

It follows from this definition in accordance with the elementary truth-table for ≡ that two propositional forms F and G are logically equivalent to one another if and only if in every possible case each has the same truth-value as the other. From examples (4) and (3) above we see that the propositional form, $p \supset q$, is logically equivalent to $\sim q \supset \sim p$ but is not logically equivalent to $q \supset p$. For

(M) $[p \supset q] \equiv [\sim q \supset \sim p]$ is a tautology, whereas
(L) $[p \supset q] \equiv [q \supset p]$ is not a tautology.

With the concept of logical equivalence at our disposal we are able to point out and justify certain ways in which some truth-functors may be defined in terms of others. For example we may define the functor v by means of $\supset$ and $\sim$ as follows:

$$p \vee q =_{df} \sim p \supset q.$$

This is an acceptable definition if and only if the truth-value of the *definiens* ($\sim p \supset q$) is in all cases the same as that of the *definiendum* ($p \vee q$). But this will be so if and only if the *definiens* and the *definiendum* are logically equivalent. That the two formulae are logically equivalent is evident from the following truth-table in which the formula: $[p \vee q] \equiv [\sim p \supset q]$ is shown to be a tautology:

p	q	$[p \vee q] \equiv [\sim p \supset q]$
1	1	$[1 \vee 1] \equiv [\sim 1 \supset 1] = 1 \equiv [0 \supset 1] = 1 \equiv 1 = 1$
1	0	$[1 \vee 0] \equiv [\sim 1 \supset 0] = 1 \equiv [0 \supset 0] = 1 \equiv 1 = 1$
0	1	$[0 \vee 1] \equiv [\sim 0 \supset 1] = 1 \equiv [1 \supset 1] = 1 \equiv 1 = 1$
0	0	$[0 \vee 0] \equiv [\sim 0 \supset 0] = 0 \equiv [1 \supset 0] = 0 \equiv 0 = 1$

The Truth-Table Method

The following definitions are also appropriate. Their testing by the method shown is left as an exercise for the reader.

$$p \cdot q =_{df} \sim [p \supset \sim q]$$
$$p \equiv q =_{df} [p \supset q] \cdot [q \supset p].$$

Since v and $.$ are both definable by means of $\sim$ and $\supset$ and $\equiv$ is definable by means of $\supset$ and $.$, it is evident that the functors v, $.$ and $\equiv$ could be dispensed with.

The five functors might, however, equally well be reduced to a different pair: either to $\sim$ and v or to $\sim$ and $.$; the reader may like to construct and to test the appropriate definitions.

9. The truth-table method applied to arguments. In the last two sections we have seen how truth-tables may be used for the purpose of determining whether or not an argument form is valid or whether or not a propositional form is tautologous. In the present section we shall show that by a formally identical method truth-tables may be used to answer the corresponding questions about arguments and propositions respectively.

Consider the argument:

(J₁) Robinson is under twenty-one or he is a graduate.

It is not the case that Robinson is under twenty-one.

Therefore he is a graduate.

This argument may be expressed as follows with the appropriate truth-functors replacing *or* and *it is not the case that*.

(J₁) Robinson is under twenty-one v he is a graduate.

$\sim$[Robinson is under twenty-one].

Therefore he is a graduate.

The Truth-Table Method

Now let us put T for the proposition *Robinson is under twenty-one* and G for the proposition *He is a graduate*. We may express (J_1) in yet a third way which we will treat as standard:

(J_1) $T \lor G$; $\sim T$; therefore G.

The corresponding implication, CI_{J_1}, for J_1 is:

(CI_{J_1}) $[[T \lor G] . \sim T] \supset G$.

J_1 is valid if and only if it exemplifies a valid argument form. This is equivalent to saying that J_1 is valid if and only if CI_{J_1} is a tautology, i.e. exemplifies a tautologous propositional form. How then can we tell whether or not CI_{J_1} exemplifies a tautologous propositional form? Before we answer this question we shall introduce the notion of a propositional form which *matches* a certain proposition.

A form F *matches* a proposition P if and only if
   (i) P can be obtained from F by consistent substitution of truth-functionally atomic propositions for all variables in F;
and (ii) F can be obtained from P by consistent substitution of variables for all truth-functionally atomic propositions in P.

Substitution is consistent when for any one symbol (proposition or variable) the same substitution is made at each occurrence. It can be seen that if a proposition P is matched by a form F then P is an exemplification of F. However, it is not necessarily true that if P exemplifies F it is matched by F. For example the proposition $[\sim T . G] \lor T$ where T and G have the meanings assigned in the last paragraph, exemplifies all the forms, among others, which we now list, but is matched only by the last two:

p, q, $p \lor q$, $[p . q] \lor r$, $[\sim p . q] \lor r$, $[\sim p . q] \lor p$, $[\sim r . p] \lor r$.

The Truth-Table Method

Let us now construct for the proposition CI_{J_1} a matching propositional form CI_J/CI_{J_1}:

CI_J/CI_{J_1} $[[p \lor q] . \sim p] \supset q$.

Since CI_J/CI_{J_1} matches CI_{J_1} we are able to construct the following joint truth-table:

CI_J/CI_{J_1} CI_{J_1}	p T	q G	$[[p \lor q] . \sim p] \supset q$ $[[T \lor G] . \sim T] \supset G$
	1	1	$[[1 \lor 1] . \sim 1] \supset 1 = [1.0] \supset 1 = 0 \supset 1 = 1$
	1	0	$[[1 \lor 0] . \sim 1] \supset 0 = [1.0] \supset 0 = 0 \supset 0 = 1$
	0	1	$[[0 \lor 1] . \sim 0] \supset 1 = [1.1] \supset 1 = 1 \supset 1 = 1$
	0	0	$[[0 \lor 0] . \sim 0] \supset 0 = [0.1] \supset 0 = 0 \supset 0 = 1$

If we delete temporarily or ignore the second line of this table we find that we have a truth-table which shows that the propositional form CI_J/CI_{J_1} is a tautology. Since CI_J/CI_{J_1} is a tautologous propositional form, CI_{J_1} which exemplifies CI_J/CI_{J_1} is a tautologous proposition. Hence J_1 is a valid argument.

What would it have meant if the form constructed to match CI_{J_1} had turned out not to be a tautology? Let us consider an example where such is the case. Take the argument:

   (K_1) $[T \lor G] . G$; therefore $\sim T$.

CI_{K_1} is $[[T \lor G] . G] \supset \sim T$. We construct in the way explained the matching form (CI_K/CI_{K_1}) $[[p \lor q] . q] \supset \sim p$ and the joint truth-table:

p T	q G	$[[p \lor q] . q] \supset \sim p$ $[[T \lor G] . G] \supset \sim T$
1	1	$[[1 \lor 1] . 1] \supset \sim 1 = [1.1] \supset \sim 1 = 1 \supset 0 = 0$
1	0	
0	1	
0	0	

The Truth-Table Method

We see from this truth-table that CI_K/CI_{K_1} can have an exemplification which is false, namely one in which p and q are both replaced by true propositions. Let e be such an exemplification. e is constructed from CI_K/CI_{K_1} by replacing p consistently by the same true proposition and q consistently by the same true proposition. But there is a one-to-one correspondence between the variables (p, q) of CI_K/CI_{K_1} and the propositions (T, G) of CI_{K_1}: wherever CI_K/CI_{K_1} has an occurrence of p CI_{K_1} has an occurrence of T and conversely, and similarly for q and G. Hence e could have been constructed equally well from CI_{K_1} by making consistent replacements for T and G. Suppose that e is constructed from CI_{K^1} by putting a for T and β for G and consider any form F of which CI_{K_1} is an exemplification. CI_{K_1} is constructed from F by the replacing of variables in F by propositional components of CI_{K^1}. Where F is CI_K/CI_{K_1}, CI_{K_1} is constructed by replacing p by T and q by G but if F is for example $[p.q] \supset r$ CI_{K_1} is constructed by putting $T \lor G$ for p, G for q and $\sim T$ for r. We can see now that whatever form F is, if it is exemplified by CI_{K_1} it must be exemplified by e; for if in the replacements for the variables of F which are required to give CI_{K_1} we put a instead of T and β instead of G we are bound to obtain e.

e then is an exemplification of any form which CI_{K_1} exemplifies. But e is false. Therefore any form which CI_{K_1} exemplifies is non-tautologous. So every truth-functional argument form exemplified by K_1 is invalid and K_1 is therefore seen to be a truth-functionally invalid argument.

The considerations which we have applied to these two examples are quite general. For any argument A_1 we can construct the corresponding implication CI_{A_1}. For any such implication we can construct a matching form CI_A/CI_{A^1} in the way described for CI_{K_1}. If CI_A/CI_{A_1} is tautologous A must be truth-functionally valid. If on

The Truth-Table Method

the other hand CI_A/CI_{A_1} is not tautologous we can see for reasons of the kind given in the discussion of K_1 that any form which CI_{A_1} exemplifies must be non-tautologous; and A_1 must therefore be truth-functionally invalid.

Arguments, as well as argument forms, may therefore be tested for validity by the truth-table method. The possibility of constructing a matching form enables us to justify our procedure but of course in any particular test the actual construction of one would be superfluous. A sufficient rule is: to test an argument A_1 for truth-functional validity or invalidity by means of truth-tables mark the truth-functionally atomic elements of A_1; treat these as if they were propositional variables and find out by means of truth-tables whether in that case A_1 would be a valid or an invalid form. If it would be valid A_1 is in fact a truth-functionally valid argument, otherwise it is invalid. For example, to test whether the argument J_1 is or is not valid all the work that is needed is the first truth-table on page 60 with the top line (the one containing p's and q's) deleted; to test K_1, all that is needed is the second table on page 60 with the top line deleted. The truth-table method of testing arguments is thus formally identical with the truth-table method of testing argument forms.

SUGGESTED READING

Quine, W. V.: *Methods of Logic*, sections 5–9.
Copi, I. M.: *Symbolic Logic*, chapter ii.

Chapter Four

THE DEDUCTIVE METHOD

1. Introduction. What we are here calling the deductive method involves the construction, in accordance with certain rules, of sequences of formulae. The formulae which make up a sequence are to be known as *lines* of the sequence. Each sequence so constructed we call a *formal deduction*. If the rules are properly chosen each deduction constructed in accordance with them is a source of information about the validity of one or more arguments or argument forms. Many different sets of rules may be adopted and each such set may be known as a *deductive system*. We will begin by setting out a set of three rules with reference to which the main principles underlying the deductive method can be explained.

Our first rule which may be called the *premiss rule* is:

(R1) Any formula may be written as a line of a sequence and marked as a premiss.

Rules (R2) and (R3) are known respectively as *modus tollens* and *modus tollendo ponens*.

(R2) If formulae $P \supset Q$ and $\sim Q$ occur earlier in the sequence then $\sim P$ may be written as a new line.

(R3) If formulae $P \lor Q$ and $\sim P$ occur earlier in the sequence then Q may be written as a new line.

The letters P, Q which occur in these rules represent either propositions or propositional forms. Thus a sequence constructed in accordance with them may be either a sequence of propositions or a sequence of propositional forms. Suppose it is a sequence of propositions; then if we take any line l, the fact that we have constructed the sequence down to l enables us to know (i) that if all the lines used as premisses are true then l is also true, and (ii) that the argument which has for its premisses all the lines which have been marked as premisses in our sequence and as conclusion the line l is a valid argument.

Suppose on the other hand that the sequence is a sequence of propositional forms; then the fact that we have constructed the sequence down to line l in accordance with the rules enables us to know that the argument form which has for its premisses all the lines which have been marked as premisses and for its conclusion the line l is a valid argument form.

We will now give examples of deductions constructed in accordance with these rules and show that the general statements which we have just made are borne out.

First of all we construct a sequence S consisting of propositional forms:

$$S$$

$$\overline{(1)}\ p \supset q$$

$$\overline{(2)}\ p \vee r$$

$$\overline{(3)}\ \sim q$$

$$(4)\ \sim p$$

$$(5)\ r$$

The first three lines of S are in accordance with the premiss rule, the sign $\overline{\vert}$ being used to mark a premiss. Line (4) is in accordance with R2 since $p \supset q$ and $\sim q$ occur as earlier lines. Line (5) is in accordance with R3 since $p \vee r$ and $\sim p$ occur as earlier lines.

Now according to what we said above this sequence

64

enables us to know that the argument form which has (1), (2) and (3) for premisses and line (4) or line (5) of the sequence for conclusion is valid. How is this assertion justified? Let us consider the case where l is line (5). We want to show that the fact that the sequence S has been constructed in accordance with the rules enables us to know that the argument form:

(A)   (1) $p \supset q$ (2) $p \lor r$ (3) $\sim q$; therefore r

is valid.

Let us write H_p for the proposition *Homer wrote the Aeneid*,

L_q for the proposition *Homer wrote in Latin*,

and G_r for the proposition *Virgil wrote the Aeneid*.

Now we set beside the sequence S another sequence S_1 in which all the lines are propositions:

S	S_1
(1) $p \supset q$	(1) $H_p \supset L_q$
(2) $p \lor r$	(2) $H_p \lor G_r$
(3) $\sim q$	(3) $\sim L_q$
(4) $\sim p$	(4) $\sim H_p$
(5) r	(5) G_r

We consider first this second sequence S_1. S_1 is, of course, constructed in accordance with our three rules. We will show first of all that if the three premiss lines (1), (2) and (3) are true then line (5) is true also. We set out our reasoning in the form of an explanatory supplement (ES) for each of the lines (4) and (5).

ES4   (*a*) If all premisses are true then lines (1), (2) and (3) are true;

(*b*) If lines (1), (2) and (3) are true then lines (1) and (3) are true;

> (*c*) If lines (1) and (3) are true then line (4) is true;

therefore (*d*) If all the premisses are true then line (4) is true.

ES5   (*a*) If all premisses are true then lines (1), (2), (3) and (4) are true;

> (*b*) If lines (1), (2), (3) and (4) are true then lines (2) and (4) are true;

> (*c*) If lines (2) and (4) are true then line (5) is true;

therefore (*d*) If all premisses are true then line (5) is true.

We can see that in each ES the crucial assertion is (*c*). ES4(*a*) is obvious and ES5(*a*) follows from ES4(*a*) and ES4(*d*). (*b*) is obvious in each stage and (*d*) obviously follows from (*a*), (*b*) and (*c*). Consider (*c*) in ES4. Here we have references to line (4) $\sim H_p$ and to the two lines (1) $H_p \supset L_q$ and (3) $\sim L_q$ in virtue of the earlier occurrence of which R2 allowed us to write line (4) in the sequence S_1. In the cases of rules R2 and R3 let us call the lines which have to occur earlier if the rule is to apply the *antecedent* lines and the line which may be written the *consequent* line. Since the rule R2 corresponds to the valid argument form: $p \supset q$, $\sim q$; *therefore* $\sim p$, we can see that in any case in which it is applied to propositions if the antecedent lines are true the consequent line must be true also. In the present case, therefore, if the antecedent lines (1) and (3) are true the consequent line (4) is also true. Hence the assertion ES4(*c*) is justified.

ES5(*c*) concerns lines (2) and (4) which are used as antecedent lines in the application of R3 and line (5) which is used as consequent line. Now R3 corresponds to the valid argument form: $p \lor q$, $\sim p$; *therefore* q. Hence in any case in which it is applied to propositions, if the antecedent lines are true the consequent line is true also.

The Deductive Method

ES5(c) which is the assertion that if (2) and (4) are true, (5) is true also is justified.

ES4(c) and ES5(c) are therefore both justified assertions. We are therefore justified in asserting ES4(d) and ES5(d) and so have achieved our first object of showing that if the premiss lines (1), (2) and (3) of S_1 are true, line (5) is also true.

We now note that the argument:

(A$_1$) (1) $H_p \supset L_q$; (2) $H_p \lor G_r$; (3) $\sim L_q$; therefore (5) G_r

is an exemplification of the argument form:

(A) (1) $p \supset q$, (2) $p \lor r$, (3) $\sim q$; therefore r

being constructed from it by the replacing of p by H_p, q by L_q and r by G_r. So what we have just shown is equivalent to showing that in this particular exemplification A$_1$ of the form A it is not the case that all the premisses are true and the conclusion is false. Now consider any other exemplification A$_i$ of the form A. It must have a p-replacement (corresponding to H_p), a q-replacement (corresponding to L_q) and an r-replacement (corresponding to G_r). Whatever these replacements may be it is evident that we can make them throughout the sequence S (or if we like throughout the sequence S_1) and that if we do so the resulting sequence S_i will still be constructed entirely in accordance with the rules R1, R2 and R3. Further, we should be able to show with regard to this sequence S_i, exactly in the same way as we did with regard to the sequence S_1, that if the premiss lines (1), (2) and (3) are true the line (5) is true also; in other words that in the exemplification A$_i$ of A it is not the case that all the premisses are true and the conclusion is false. But since A$_i$ may be any exemplification whatever it is evident that no exemplification of A can have true premisses and false conclusion. Hence A must be a valid argument form.

In this way the fact that we can construct the sequence S in accordance with our three rules enables us to know

The Deductive Method

that the argument form A which has the premiss lines of S as its premisses and line (5) for its conclusion is valid.

The same thing can be shown in a similar way with regard to any line l of any sequence constructed in accordance with our rules, where all the lines are propositional forms. The argument form which has for its premisses all the premiss lines of the sequence and for conclusion the line l is a valid argument form.

Suppose now that we construct a sequence T_i which consists not of propositional forms but of propositions. What are we able to know as a result of having constructed T_i? In the first place we know that if all the premiss lines of T_i are true any line l of T_i is true also. We can know this by the same sort of reasoning which we had before us when considering by itself the sequence S_1. However, we can know something further, namely that any argument which has for premisses all the premiss lines of T_i and for conclusion any line l of T_i is a valid argument. We will now show that this last assertion is justified.

Suppose that T_i is a sequence which has for premiss lines propositions $P_i, Q_i, ..., R_i$. Let L_i be any line of T_i. We want to show that the argument:

(B_i) $P_i, Q_i, ..., R_i$; therefore L_i

is valid. By the method shown in section 9 of the last chapter we construct a matching form:

(B/B_i) $P, Q, ..., R$; therefore L.

To do this we replace the distinct atomic propositions which are components of the propositions of B_i by distinct variables. Now by replacing these same propositions by these same variables throughout the whole sequence T_i down to the line L_i we obtain a sequence of propositional forms. Clearly this sequence is correctly constructed in accordance with our three rules and has for its premiss lines the premisses of B/B_i and for its last line the conclusion of B/B_i. Since a correctly constructed sequence with these properties is possible we know from our con-

The Deductive Method

siderations a little earlier that the argument form B/B_i must be valid. But the argument B_i is an exemplification of its matching argument form B/B_i. Hence, since B/B_i is a valid form, B_i is a valid argument.

Hitherto we have used the term *valid* in connexion with both arguments and argument forms. It is convenient now to define a third use of the term, this time in connexion with rules of deduction. As we have seen a rule is applicable either to propositions or to propositional forms. Let us say that a rule is valid if it is such that

(i) when it is correctly applied to propositions it is always the case that if the antecedent propositions are true the consequent proposition is true also;

and

(ii) when it is correctly applied to propositional forms there is no possible exemplification of antecedents and consequent such that the antecedents are all true and the consequent is false.

The rules R1, R2 and R3 are all valid in this sense. We have explained how a deduction constructed in accordance with these rules serves certain purposes; for example it enables us to know that a certain argument or argument form is valid. It should be clear, however, that the same considerations may be extended to cover deductions constructed in accordance with any rules which are valid in the sense defined. A deduction constructed in accordance with any system of valid rules will have the properties which we have been discussing.

2. A deductive system; initial rules. The set of three rules R1, R2 and R3 which we used for illustrative purposes in the last section would by themselves enable us to prove the validity of only a very small proportion

The Deductive Method

of all possible truth-functionally valid arguments and argument forms. In this section we will set forth a more extensive and useful system of valid rules and explain in some detail how these are to be applied. This system which is derived with modifications largely from one set forth by I. M. Copi[1] is an example of what is called a *natural deductive system*.

Our first rule is again the rule R1. We call this by the name *Premiss Rule* and use Pr as an abbreviation.

> Premiss Rule (Pr). Any formula (i.e. any proposition or truth-functional propositional form) may be written as a line of a deduction and marked as a premiss.

Of the remaining rules some are on the following pattern:

> If P, Q, R, ... are propositions (propositional forms) then if W_1, W_2, ... occur as lines of a deduction Z may be written as a new line.

Here W_1, W_2, ..., Z represent propositions (propositional forms) which have some or all of P, Q, R, ... as components. An example of a rule of this sort is the second rule of our system:

> *Modus ponens* (MP). If P and Q are propositions (propositional forms) then if $P \supset Q$ and P are lines of a deduction Q may be written as a new line.

Normally we will not write a rule of this sort out in full but will abbreviate it in the form:

$$W_1, W_2, ... \rightarrow Z.$$

Thus we write the rule *Modus ponens:*
Modus ponens (MP). $P \supset Q, P \rightarrow Q$.

Other rules in this group are the following:

---

[1] See *Suggested Reading*, p. 95.

70

The Deductive Method

Modus tollens (MT). $P \supset Q, \sim Q \to \sim P$.
Hypothetical Syllogism (HS). $P \supset Q, Q \supset R \to P \supset R$.
Modus tollendo ponens (MTP). $P \lor Q, \sim P \to Q$.
Addition (Add). $P \to P \lor Q$.
Simplification (Simp). $P.Q \to P; P.Q \to Q$.
Conjunction (Conj). $P, Q \to P.Q$.

It can easily be checked that each of these rules corresponds directly to a valid argument form and so is a valid rule in the sense explained in the last section. For example the rule HS corresponds to the valid argument form:

$$p \supset q, q \supset r; \text{ therefore } p \supset r.$$

We give now examples of deductions constructed in accordance with this group of rules. A deduction, as we explained, is a sequence of formulae, each formula being known as a line of the deduction or a main line or a deduction line. Before each deduction line we insert an indented line called a *justification line*. The justification line is not itself a part of the deduction, but serves to indicate by what authority the immediately following deduction line is written and thus enables anyone to check fairly quickly whether or not the deduction is properly constructed. The justification line indicates in each case (i) the antecedent lines, if any, for the rule which is being employed; (ii) the rule itself, and (iii) the number of the line which is to be written.

Example 1

Let *J* represent the proposition: *The secretary's action was justified.*

Let *A* represent the proposition: *The secretary's action was authorized by the committee.*

Let *M* represent the proposition: *There is an entry in the Minute Book.*

We may construct the following deduction:

71

$$
\begin{array}{l}
\quad\quad \text{Pr} \times 1 \\
(1)\ \ J \supset A \\
\quad\quad \text{Pr} \times 2 \\
(2)\ \ A \supset M \\
\quad\quad \text{Pr} \times 3 \\
(3)\ \ \sim M \\
\quad\quad 1,\ 2,\ \text{HS} \times 4 \\
(4)\ \ J \supset M \\
\quad\quad 4,\ 3,\ \text{MT} \times 5 \\
(5)\ \ \sim J
\end{array}
$$

The first justification line indicates that the premiss rule is used. The sign × may be read as *yields*. The justification line for line 4 indicates that lines (1) and (2) by the rule of hypothetical syllogism yield line (4), i.e. that from lines (1) and (2) by means of this rule we may obtain line (4). The deduction as a whole shows that the argument with lines (1), (2), (3) as premisses and with line (5) as conclusion is valid; as is indeed implied in this assertion it shows also that if lines (1), (2), (3) are all true then line (5) is true also; this in its turn implies that if we have begun knowing that (1), (2), (3) are all true we are able to know, having constructed the deduction, that line (5) is true also.

It is worth remarking that the sequence of five lines which we have been treating as a deduction ending in line (5) really comprises in itself five deductions, namely those ending in lines (1), (2), (3), (4) and (5) respectively. Only the last two, of course, are of any particular interest. In virtue of the second last deduction we are able to say that the argument form with lines (1), (2) and (3) as premisses and line (4) as conclusion is valid. However, when we set out a sequence of lines as a deduction we are normally interested in it as a deduction consisting of all the lines shown, in our present example of lines (1), (2), (3), (4) and (5).

The Deductive Method

Example 2

This example illustrates, among other things, the use of the rules of addition, simplification and conjunction.

$$\text{Pr} \times 1$$
(1) $[p . \sim q] \supset r$

$$\text{Pr} \times 2$$
(2) $[rvs] \supset t$

$$\text{Pr} \times 3$$
(3) $p . [q \supset u]$

$$\text{Pr} \times 4$$
(4) $\sim u$

$$3 \text{ Simp} \times 5$$
(5) p

$$3 \text{ Simp} \times 6$$
(6) $q \supset u$

$$6, 4, \text{MT} \times 7$$
(7) $\sim q$

$$5, 7, \text{Conj.} \times 8$$
(8) $p . \sim q$

$$1, 8, \text{MP} \times 9$$
(9) r

$$9, \text{add} \times 10$$
(10) rvs

$$2, 10, \text{MP} \times 11$$
(11) t

This deduction shows that the argument form with lines (1), (2), (3) and (4) as premises and line (11) as conclusion is a valid argument form.

3. The rule of interchange. It remains to introduce two rules which are rather more complicated than those dealt with in section 2. We will deal in this section with

The Deductive Method

the rule of interchange and in section 4 with the rule of conditional proof.

Let P, K and K' be propositions such that (i) K occurs at one or more places as a truth-functional component of P, (ii) K is logically equivalent to K'. The term *component* has hitherto been used in such a way that K is a component of P if P can be constructed by means of truth functors out of a number of propositions of which K is one; in (i) and hereafter the term is used in an extended sense which covers also the case where K is identical with P. Now it follows from the definition of logical equivalence that since K and K' are logically equivalent they must have the same truth-value: either both are true or both are false. So if we replace K in P by K' and call the resulting proposition P' the truth-value of P' will be the same as that of P. For the truth-value of P depends, so far as its truth-functional components are concerned, solely on their truth-values and the operation described consists in substituting for a truth-functional component another which has the same truth-value. The result is the same whether K is identical with P or with a proper part of P and whether the substitution is made at all occurrences of K in P or at some only but not all; in all cases P' must have the same truth value as P. Hence if P is true P' must be true also. Further, since this holds whatever propositions P, K and K' may be, provided only that the stated conditions are satisfied, it follows that if P, K and K' are taken to be not propositions but propositional forms the argument form $P \therefore P'$ must be valid. Consequently, whether P, K and K' are propositions or propositional forms so long as the stated conditions are satisfied we may write P' as a line of deduction where P has occurred as an earlier line. Let us use the sign $P^{K:K'}$ to denote a proposition or propositional form which is the result of substituting K' for K at one or more occurrences in P. Let us say also that one propositional form or argument

The Deductive Method

form is a *specification* of another if every exemplification of the first is an exemplification of the second. We may now express the *rule of interchange* as follows:

Rule of interchange (Int)
$P \to P^{K;K'}$, where either $K \equiv K'$ or $K' \equiv K$ is an

exemplification or a specification of one of the logical equivalences in the following list.

Logical equivalences

In the list which follows the main functor in each formula is indicated by spacing instead of by bracketing. (The equivalences in the left-hand column are by themselves sufficient, but in some cases useful auxiliaries are shown on the right.)

I. De Morgan's Laws
$$\sim[p.q] \equiv \sim p \vee \sim q$$
$$\sim[p \vee q] \equiv \sim p . \sim q$$

II. Laws of Commutation
$$p.q \equiv q.p$$
$$p \vee q \equiv q \vee p$$

III. Laws of Association
$$p.[q.r] \equiv [p.q].r$$
$$p \vee [q \vee r] \equiv [p \vee q] \vee r$$

IV. Laws of Distribution
$$p.[q \vee r] \equiv [p.q] \vee [p.r]$$
$$p \vee [q.r] \equiv [p \vee q].[p \vee r]$$

V. Law of Double Negation
$$p \equiv \sim \sim p$$

VI. Law of Exportation and Importation
$$[p.q] \supset r \equiv p \supset [q \supset r] \qquad p \supset [q \supset r] \equiv q \supset [p \supset r]$$

VII. Law of Transposition
$$p \supset q \equiv \sim q \supset \sim p \qquad p \supset \sim q \equiv q \supset \sim p$$
$$\sim p \supset q \equiv \sim q \supset p$$

VIII.

$$p \lor q \equiv \sim p \supset q \qquad\qquad p \supset q \equiv \sim p \lor q$$
$$p \supset q \equiv \sim [p . \sim q]$$

IX.

$$p \equiv q \equiv [p \supset q].[q \supset p] \qquad p \equiv q \equiv [p.q] \lor [\sim p. \sim q]$$

X.

$$p \lor p \equiv p \qquad\qquad \sim p \supset p \equiv p$$
$$p \supset \sim p \equiv \sim p$$

XI. $\quad p . p \equiv p$

When the Rule of Interchange is used the justification line is constructed thus: first we give the number of the formula which is our antecedent; then we write *Int* and in brackets after it the number of the equivalence to be referred to; e.g. if we have a line: (m) $p \supset [q \lor r]$, and we apply the rule of interchange to give us the line: (n) $p \supset \sim \sim (q \lor r)$, our justification line will be: *m, Int (V)* × *n*.

We now construct a deduction which contains some examples of the use of the rule of interchange.

Example:

The argument form:

$\sim p \lor q$, $[p \supset r] \supset \sim [s \lor t]$, $q \supset r$; therefore $\sim t$

is shown to be valid by the following deduction:

$$\begin{array}{l}
\quad\quad \text{Pr} \times 1 \\
(1) \ \sim p \lor q \\
\quad\quad \text{Pr} \times 2 \\
(2) \ [p \supset r] \supset \sim [s \lor t) \\
\quad\quad \text{Pr} \times 3 \\
(3) \ q \supset r \\
\quad\quad 1, \text{int (VIII)} \times 4 \\
(4) \ p \supset q \\
\quad\quad 4, 3, \text{HS} \times 5 \\
(5) \ p \supset r \\
\quad\quad 2, \text{int (I)} \times 6
\end{array}$$

The Deductive Method

(6) $[p \supset r] \supset [\sim s . \sim t]$
    6, 5, MP × 7

(7) $\sim s . \sim t$
    7, int (II) × 8

(8) $\sim t . \sim s$
    8, simp × 9

(9) $\sim t$

The rule of interchange is used on three occasions. At step 4 P is $\sim pvq$ and K is identical with P. K' is $p \supset q$ and the equivalence $K' \equiv K$ is identical with and hence a specification of the equivalence VIII: $[p \supset q] \equiv [\sim pvq]$. At step 6 P is the formula $[p \supset r] \supset \sim [svt]$, K is $\sim [svt]$, K' is $\sim s . \sim t$ and the equivalence $\sim [svt] \equiv [\sim s . \sim t]$ is a specification of one of De Morgan's laws. At step 8 P is $\sim s . \sim t$, K is identical with P, and K' is $\sim t . \sim s$ and the equivalence $[\sim s . \sim t] \equiv [\sim t . \sim s]$ is a specification of one of the laws of commutation. It should be noted, however, that we could have gone directly from 7 to 9 by using the second form of the rule *simp*.

4. The Rule of Conditional Proof. Let P_1, P_2, ..., P_{n-1}, P_n and Q by any propositions such that:

(Z) P_1, P_2, ..., P_{n-1}, P_n; therefore Q,

is a valid argument. We assume for the present that n is greater than 1, i.e. that Z has at least two premisses. Since Z is a valid argument it follows—see chapter iii, section 6—that the corresponding conditional:

(CI$_z$) $[P_1 . P_2 . \ldots . P_{n-1} . P_n] \supset Q$,

is a tautology. The antecedent of CI$_z$ may be written as $[[P_1 . P_2 . \ldots . P_{n-1}] . P_n]$. Let us put S for $[P_1 . P_2 . \ldots . P_{n-1}]$ CI$_z$ may now be written:

(CI$_z$) $[S . P_n] \supset Q$.

Since CI$_z$ is a tautology the logically equivalent proposition:

77

The Deductive Method

$$S \supset [P_n \supset Q],$$

must be a tautology too. Hence the argument:

$$S; \text{ therefore } P_n \supset Q,$$

is valid also; i.e. the argument:

$$[P_1 . P_2 . \dots . P_{n-1}]; \text{ therefore } P_n \supset Q,$$

is valid and it obviously follows that the argument:

$$(Z') \quad P_1, P_2, \dots, P_{n-1}; \text{ therefore } P_n \supset Q,$$

is valid too. We see then that if Z is a valid argument Z' must be valid also. It follows that if we can construct a deduction with premisses $P_1, P_2, \dots, P_{n-1}, P_n$ and conclusion Q then the argument Z' must be valid. Further, all that has been said in this section will still hold good if we take $P_1, P_2, \dots, P_{n-1}, P_n$ and Q to be propositional forms and substitute *argument form* for *argument* throughout. We are, therefore, entitled to use the following rule.

If $P_1, P_2, \dots, P_n, Q$ are propositions (propositional forms) then if
  (i) $P_1, P_2, \dots, P_{n-1}$ occur as lines in a deduction D and
  (ii) a subsidiary deduction can be constructed with premisses
          $P_1, P_2, \dots, P_{n-1}, P_n$ and last line Q,
then we may write as a new line of D the formula $P_n \supset Q$.

We introduce now the sign ⊢ and define it as follows:

$P_1, P_2, \dots, P_n \vdash Q$ means that there is a deduction (or a deduction can be constructed) with premisses $P_1, P_2, \dots, P_n$ and last line Q.

Accordingly we may express the rule of conditional proof in the following abbreviated form:

The Deductive Method

Conditional Proof (CP)

$P_1, P_2, ..., P_{n-1}, (P_1, P_2, ..., P_{n-1}, P_n \vdash Q) \rightarrow P_n \supset Q.$

We have now to see how this rule may be used. We give first a very simple example to make sure that its terms are understood. First, we construct a deduction with premisses $p \supset r$ and $p.q$ and last line r as follows:

$$
\begin{array}{l}
\quad\quad \text{Pr} \times 1 \\
(1)\ p \supset r \\
\quad\quad \text{Pr} \times 2 \\
(2)\ p.q \\
\quad\quad 2\ \text{simp} \times 3 \\
(3)\ p \\
\quad\quad 1, 3\ \text{MP} \times 4 \\
(4)\ r
\end{array}
$$

Let us call this deduction α. Given α we may now construct the following deduction involving two lines:

$$
\begin{array}{l}
\quad\quad \text{Pr} \times 1 \\
(1)\ \text{p} \supset r \\
\quad\quad 1, \alpha, \text{CP} \times 2 \\
(2)\ [p.q] \supset r.
\end{array}
$$

We call this deduction β. Let us explain the final step of β with reference to our account of the rule CP. In the case where $n=2$ the rule would read:

$$P_1, (P_1, P_2 \vdash Q) \rightarrow P_2 \supset Q.$$

In our example P_1 is the propositional form $p \supset r$ and P_2 is $p.q$. Q is r. In the deduction α we have a deduction from $p \supset r, p.q$ to r, i.e. $p \supset r, p.q \vdash r$. Hence in accordance with the rule CP if we have a deduction in which $p \supset r$ occurs as a line we may write as a later line $[p.q] \supset r$ $(=P_2 \supset Q)$. This is what we have done in the deduction β. However, the rather informal way of using the rule of conditional proof which has been adopted in this illustration would be unsatisfactory in cases involving any

79

The Deductive Method

degree of complexity. It can be replaced by a more elegant method which will now be explained.

5. Application of the rule of conditional proof.
Each use of the rule of conditional proof requires the existence of a subsidiary deduction which has all the premisses of the main deduction and one additional premiss. We shall describe a method by which subsidiary deductions may be incorporated in a single column sequence with the main deduction. However, I think that this may be more easily understood if it is shown as a development of a method in which subsidiary deductions are set out in separate columns to the right of the main deduction. We will explain this method by means of an example. We set out below a main deduction in column 1 with subsidiary deductions in columns 2 and 3:

Column 1	*Column* 2	*Column* 3
$Pr \times 1$	$Pr \times 1$	$Pr \times 1$
(1) $[p \vee q] \supset r$	(1) $[p \vee q] \supset r$	(1) $[p \vee q] \supset r$
	$Pr \times 2$	$Pr \times 2$
	(2) $s \supset p$	(2) $s \supset p$
		$Pr \times 3$
		(3) s
		2, 3, $MP \times 4$
		(4) p
		4 add $\times 5$
		(5) $p \vee q$
		1, 5 $\times 6$
		(6) r
	1, 2 (1, 2,	
	3 $\vdash$ 6) $\times 7$	
	(7) $s \supset r$	
1 (1, 2 $\vdash$ 7) $\times 8$		
(8) $[s \supset p] \supset [s \supset r]$		

80

The Deductive Method

To understand these deductions we begin with column 3. Here we have a straightforward deduction of line (6) from premisses (1), (2) and (3) without any use of CP. In virtue of the existence of this deduction (1, 2, 3 ⊢ 6) we are able in column 2 where we have (1) and (2) but not (3) for premisses to apply CP and derive the line (7) which is $s \supset r$ or $3 \supset 6$. We have then in column 2 a deduction from premisses (1) and (2) of conclusion (7), i.e. (1, 2 ⊢ 7). In virtue of the existence of this deduction we are entitled in column 1, where we have the sole premiss (1), to apply CP and derive the line (8) which is $[s \supset p] \supset [s \supset r]$ or $2 \supset 7$. In column 1, our main deduction, we now have a deduction from premiss (1) of conclusion (8). The argument form:

(A) $[p \lor q] \supset r$; therefore $[s \supset p] \supset [s \supset r]$,

is thus shown to be valid.

In constructing these deductions, as distinct from understanding them when constructed, we begin not with column 3 but at the top of column 1. Let us suppose that we have the argument form (A) in mind and wish to show that it is valid. We start by writing down $[p \lor q] \supset r$ as a premiss, line (1), of our deduction. No immediately useful formula appears to be derivable from (1) alone; however, the conclusion which we are seeking to obtain is here an implication and when this is so CP may often be profitably used; accordingly we begin in column 2 a subsidiary deduction in which we again have (1) for a premiss but add a subsidiary premiss (2). This is the formula $s \supset p$ which is the antecedent of the conclusion which we are trying to obtain in column (1). Our object in the column 2 deduction is to obtain from premisses (1) and (2) the conclusion $s \supset r$, which is the consequent of the conclusion which we are trying to obtain in column (1). However, from (1) and (2) no obviously useful consequence immediately follows; we therefore embark on yet another subsidiary deduction in

The Deductive Method

column 3 where we add to premisses (1) and (2) the subsidiary premiss (3). (3), it will be noted, is *s*, the antecedent of the conclusion which we are seeking to obtain in column 2. From column 3 we find that from premisses (1), (2), (3) the conclusion *r* can be obtained. The construction of the deduction from this point on has already been explained.

It may be useful here to remark that adding a subsidiary premiss and so beginning a subsidiary deduction in this formal deduction corresponds to making a supposition, or supposing something, in informal reasoning and then considering what would follow from the supposition. In informal reasoning if we make a supposition *S* and find that a proposition *T* follows we infer that the proposition *If S then T* is a consequence of our original premisses; this inference corresponds exactly to the application of the rule of conditional proof. We may indeed, if we prefer it, refer to the premiss rule as the *supposition rule* and instead of writing in our justification line, e.g. *Pr × 1* we might write simply *Suppose 1*. However, this possible variant will not be used in this book.

We must now show how these three deductions may be arranged more conveniently in a single column. First, however, let us observe that we might, if we had so wished, have extended the column 2 deduction beyond line (7); for example by an application of the rule of interchange we could have added a line (7a) $\sim r \supset \sim s$. Any such extension, however, unless it involves another use of CP may be made only in virtue of lines which occur in column 2 itself; thus (7a) is legitimate since it results from the application of *Int* to (7) which is a line of column 2; but it would not be legitimate to use, e.g. (1) and (5) and from these by MP to derive a line (7b) *r*; for (5) is a line of the column 3 deduction but not a line of the column 2 deduction. Formally such a move is clearly not allowable. It may be helpful, however, if we draw attention to

82

The Deductive Method

the reasons underlying the formal restriction. One essential thing about the column 2 deduction is that any line which occurs in it must be derivable ultimately from premisses (1) and (2) alone. If in extending the column 2 deduction we were to make use of a line which occurs in column 3 but not in column 2 we should be doing something which was not legitimate; for, since column 3 contains an additional premiss (3), we have no guarantee that any line occurring in column 3 but not in column 2 is derivable from premisses (1) and (2) alone.

For similar reasons if we wish to extend our column 1 deduction beyond line (8) we must refer, in the application of our rules (other than CP), only to lines which occur in column 1 itself; and in general in extending a deduction by any rule other than CP we must not make use of any line which occurs only in some subsidiary deduction but not in the deduction itself. Indeed this principle applies to the use of CP too; in justifying the use of CP in a certain main deduction D we refer to certain lines and to a certain subsidiary deduction S. S, of course, is not part of D but the lines to which we refer must be part of D itself. For example if in a main deduction D we write the justification line:

$$1, 3, (1, 3, 5 \vdash 9) \times 10$$

the lines 1 and 3 must be lines of D itself, though of course the deduction $(1, 3, 5 \vdash 9)$ is a subsidiary deduction and not part of D itself.

An advantage of setting out our deductions in distinct columns is that these restrictions and the reasons for them are graphically obvious. If we wish to present a set of deductions, main and subsidiary together, in a single column we must adopt some means whereby it can be seen at a glance which lines belong to a deduction D itself and which only to a deduction subsidiary to D. Various good devices are in use. Here we shall use a method involving boundary lines. We will explain this method in the course of the account which we now proceed to give of how a set of related deductions may be arranged in one column. We refer to the example which we have had earlier. The first step is to take the extreme right-hand deduction, i.e. column 3, and delete from it any formula which also occurs at the same level in column 2. This means that we delete lines (1) and (2) from column 3. Next we draw a horizontal line immediately under the conclusion (6) of the column 3 deduction, and we join the left-hand end of this line to the downward stroke of the sign marking the subsidiary premiss (3). This means that that part of the column 3 sub-

Column 1	*Column 2*	*Column 3*

$\text{Pr} \times 1$

(1) $[p \lor q] \supset r$

$\text{Pr} \times 2$

(2) $s \supset p$

$\text{Pr} \times 3$

(3) s

2, 3, MP $\times$ 4

(4) p

4 add $\times$ 5

(5) $p \lor q$

1, 5, MP $\times$ 6

(6) r

1, 2, (1, 2, 3 ⊢ 6) $\times$ 7

(7) $s \supset r$

1, (1, 2 ⊢ 7) $\times$ 8

(8) $[s \supset p] \supset [s \supset r]$

84

The Deductive Method

sidiary deduction which does not duplicate column 2 is now partially bounded by a line. We now repeat this procedure in respect of the column 2 deduction. We delete any lines of column 2 which duplicate lines of column 1; we draw a horizontal line under the conclusion (7) of the column 2 deduction and join this line up to the sign marking the subsidiary premiss (2). What we now have is shown on page 84:

To obtain a single column arrangement we now move column 3 horizontally across into the vacant place in column 2; this new column 2 in its turn is moved across into the vacant place in column 1. In the end we have the arrangement shown below with subsidiary deductions separated from main deductions by boundary lines.

$\text{Pr} \times 1$
(1) $[p \lor q] \supset r$

$\text{Pr} \times 2$
(2) $s \supset p$

$\text{Pr} \times 3$
(3) s
2, 3, MP $\times$ 4
(4) p
4, add $\times$ 5
(5) $p \lor q$
1, 5, MP $\times$ 6
(6) r

1, 2, (1, 2, 3 $\vdash$ 6) $\times$ 7
(7) $s \supset r$

1, (1, 2 $\vdash$ 7) $\times$ 8
(8) $[s \supset p] \supset [s \supset r]$

The Deductive Method

What we have here is really a set of nested deductions, one main deduction, the original column 1, and two subsidiary ones. We shall refer to the whole thing, however, as 'a deduction'. It is intended of course that every deduction should be constructed from the beginning in a single column, the three-column arrangement having been used here merely for explanatory purposes. Every time that the rule of conditional proof is used we mark off with a boundary line the subsidiary deduction on which it depends or rather that part of the subsidiary deduction which is not also part of the main deduction. When CP is used the subsidiary premiss is said to be discharged and the line which we draw under the conclusion of the subsidiary deduction may be called a *discharge line*. A deduction is said to be a deduction from its undischarged premisses to its last line. In our example we have a deduction from (1) which is the only undischarged premiss to (8). However, if we had stopped immediately after obtaining line (7) we should have had a deduction from (1) and (2) to (7); and if we had stopped immediately after (6) we should have had a deduction from (1), (2) and (3) to (6).

Our deduction may of course be extended in numerous different ways. We will describe one possible extension which involves the introduction of a new premiss into our main deduction and another use of the rule of conditional proof. Our new premiss which we bring in as line (9) is the formula: $r \supset p$. After adding this line to our main deduction we begin yet another subsidiary deduction with subsidiary premiss $s \supset r$. We show this subsidiary deduction first of all in a distinct column 4.

The Deductive Method

Column 1	Column 4

Column 1

```
      Pr × 1
(1) [pvq]⊃r

      Pr × 2
(2) s⊃p

      Pr × 3
(3) s
      2, 3, MP × 4
(4) p
      4, add × 5
(5) pvq
      1, 5, MP × 6
(6) r

      1, 2, (1, 2, 3 ⊢ 6)
      × 7
(7) s⊃r

      1, (1, 2 ⊢ 7) × 8
(8) [s⊃p]⊃[s⊃r]

      Pr × 9
(9) r⊃p
```

Column 4

```
      Pr × 1
(1) [pvq]⊃r

      Pr × 2
(2) s⊃p

      Pr × 3
(3) s
      2, 3, MP × 4
(4) p
      4, add × 5
(5) pvq
      1, 5, MP × 6
(6) r

      1, 2, (1, 2, 3 ⊢ 6)
      × 7
(7) s⊃r

      1, (1, 2 ⊢ 7) × 8
(8) [s⊃p]⊃[s⊃r]

      Pr × 9
(9) r⊃p

      Pr × 10
(10) s⊃r
      10, 9, HS × 11
(11) s⊃p
```

In column 4 we have a subsidiary deduction from (1), (9) and (10) to (11). Since (1) and (9) occur in the main column 1 deduction we may apply the rule of conditional

The Deductive Method

proof and add line (12) $[s \supset r] \supset [s \supset p]$ to column 1. We show now the whole deduction arranged in one column and extended to a line (14).

$$\text{Pr} \times 1$$
(1) $[p \lor q] \supset r$

$$\text{Pr} \times 2$$
(2) $s \supset p$

$$\text{Pr} \times 3$$
(3) s

$$2, 3, \text{MP} \times 4$$
(4) p

$$4, \text{add} \times 5$$
(5) $p \lor q$

$$1, 5, \text{MP} \times 6$$
(6) r

$$1, 2, (1, 2, 3 \vdash 6) \times 7$$
(7) $s \supset r$

$$1, (1, 2 \vdash 7) \times 8$$
(8) $[s \supset p] \supset [s \supset r]$

$$\text{Pr} \times 9$$
(9) $r \supset p$

$$\text{Pr} \times 10$$
(10) $s \supset r$

$$10, 9, \text{HS} \times 11$$
(11) $s \supset p$

The Deductive Method

$$1, 9, (1, 9, 10 \vdash 11) \times 12$$
(12) $[s \supset r] \supset [s \supset p]$
$$8, 12, \text{conj} \times 13$$
(13) $[[s \supset p] \supset [s \supset r]] . [[s \supset r] \supset [s \supset p]]$
$$13 \text{ int (IX)} \times 14$$
(14) $[s \supset p] \equiv [s \supset r]$

This sequence as a whole is a deduction to its last line (14) from the undischarged premisses (1) and (9). Down to line (12) or (13) it is a deduction from the same premisses to the line in question. Down to line (11) it is a deduction to (11) from the still undischarged premisses (1), (9) and (10).

6. Use of the deductive method to prove logical truth. At the beginning of section 4 we showed that given propositions (or propositional forms) P_1, P_2, ..., P_n, Q such that (i) n is greater than 1 and (ii) the argument (or argument form)

(Z) P_1, P_2, ..., P_n; therefore Q

is valid, the argument (or argument form)

(Z') P_1, P_2, ..., P_{n-1}; therefore $P_n \supset Q$

must be valid too. This consideration justified the rule of conditional proof as it has been applied hitherto, that is in cases where, in the subsidiary deduction $(P_1, P_2, ..., P_n \vdash Q)$, n is greater than 1, or in other words where the subsidiary deduction has at least two premisses. We now turn to consider the possibility of applying the rule in the case where $n=1$, i.e. where the subsidiary deduction has a single premiss only.

Suppose that we have a deduction with a single undischarged premiss P_1 and last line Q. Suppose this to be set out in a single column, column 1.

The Deductive Method

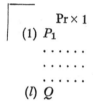

$$\text{Pr} \times 1$$
$$(1) \ P_1$$
$$\cdots\cdots$$
$$\cdots\cdots$$
$$\cdots\cdots$$
$$(l) \ Q$$

If we apply the rule of conditional proof in a way analogous to that adopted in certain cases in the previous section we should write in a distinct column 0 to the left of column 1 a new line $(l+1)$ $P_1 \supset Q$. The column 0 deduction can of course be extended in numerous different ways. We show below column 1 and column 0 with one additional line in column 0.

Column 0   *Column 1*

$$\text{Pr} \times 1$$
$$(1) \ P_1$$
$$\cdots\cdots$$
$$\cdots\cdots$$
$$\cdots\cdots$$
$$(l) \ Q$$

$$(1 \vdash l) \times l+1$$
$$(l+1) \ P_1 \supset Q$$
$$l+1 \text{ int (VII)} \times l+2$$
$$(l+2) \ \sim Q \supset \sim P_1$$

We have now a subsidiary deduction with a single premiss in column 1 and a main deduction in column 0. The striking thing about the column 0 deduction is that it contains no undischarged premisses. What, if anything, we must now ask, is the significance of a deduction without

90

The Deductive Method

premisses? This question is not difficult to answer. The line $P_1 \supset Q$ is obtained in column 0, where there are no premisses, by applying CP to the column 1 deduction which has P_1 as its single undischarged premiss and Q as its last line. Now the column 1 deduction is, we are of course assuming, properly constructed; hence the argument or argument form:

$$(Z) \quad P_1; \text{therefore } Q,$$

which has for premiss the sole undischarged premiss of that deduction and for conclusion its last line Q, must be valid. But if Z is valid the corresponding implication

$$(CI_z) \quad P_1 \supset Q,$$

which is identical with the line obtained in column 0 by CP, must be a tautology. It is clear then that any line obtained by applying CP to a subsidiary deduction which has only a single premiss must be a tautology, i.e. a truth-functional logical truth. Now it is easy to see that the *first* line of a main deduction that has no premisses can be obtained only in this way by CP. The first line then of such a deduction must be a tautology. Any subsequent line in a premiss-free deduction will be obtained either also by CP in which case it will also be a tautology or by the application to previous lines of some rule other than CP and the premiss rule. But all these rules are valid in the sense explained in section 1 (page 69), and it follows from this that when applied to tautologies they can yield only tautologies. Hence it is clear that not only the first but also every subsequent line of a deduction which has no premisses must be a tautology. We thus have a deductive method of establishing that a formula is a tautology.

The account which we have just been giving of the deductive method of establishing tautology has been given

G　　　　　　　　91

The Deductive Method

with reference to deductions in which main and subsidiary deductions are set out separately in distinct columns. This multiple-column arrangement is again adopted purely for explanatory reasons and the normal arrangement is in a single column. The schematic deduction which we have been referring to may be set out in a single column thus:

$$
\begin{array}{ll}
& \text{Pr} \times 1 \\
(1) & P_1 \\
& \cdots\cdots \\
& \cdots\cdots \\
& \cdots\cdots \\
(l) & Q
\end{array}
$$

$$
\begin{array}{ll}
& (1 \vdash l) \times l{+}1 \\
(l{+}1) & P_1 \supset Q \\
& l{+}1,\ \text{int (VII)} \times l{+}2 \\
(l{+}2) & \sim Q \supset \sim P_1
\end{array}
$$

We must now express the principle of our method in such a way that it can be easily applied to a single column deduction such as this. First we introduce the expression *undischarged at line n*. A premiss may be said to be undischarged at line n in a deduction if (i) it occurs in the deduction at or before line n and (ii) it is discharged, if at all, only after line n. Our principle may now be stated: a formula is a tautology if it occurs in a deduction as a line at which there is no premiss undischarged.

This method may now be illustrated. If we take the deduction on page 85 down to line (8) and at that point by applying CP discharge the only undischarged premiss we have the following:

92

Pr × 1
(1) $[p \lor q] \supset r$

      Pr × 2
  (2) $s \supset p$

      Pr × 3
    (3) s
      2, 3, MP × 4
    (4) p
      4 add × 5
    (5) $p \lor q$
      1, 5, MP × 6
    (6) r

      1, 2, (1, 2, 3 ⊢ 6) × 7
  (7) $s \supset r$

    1, (1, 2 ⊢ 7) × 8
  (8) $[s \supset p] \supset [s \supset r]$

(1 ⊢ 8) × 9
(9) $[[p \lor q] \supset r] \supset [[s \supset p] \supset [s \supset r]]$

At line (9) there is no undischarged premiss. Hence (9) $[[p \lor q] \supset r] \supset [[s \supset p] \supset [s \supset r]]$ is a tautology.

7. Indirect deductions; *reductio ad absurdum* deductions. Hitherto the rule of conditional proof has been used in the following way: where we have wanted to obtain a formula, $P \supset Q$, which is an implication we have introduced P as a premiss and have then attempted to derive Q. If we succeed in doing this we discharge the premiss P and use CP to obtain $P \supset Q$. However, we may also use CP where our main object is to obtain some

The Deductive Method

formula which is not an implication. Suppose that we wish to obtain a formula P. We may attempt to do this by first of all introducing $\sim P$ as a premiss; if we are then able to deduce P we may apply CP to obtain $\sim P \supset P$. But $\sim P \supset P$ is logically equivalent to P (Equivalence X); so we may use the rule of interchange to obtain P. Similarly of course if we wish to obtain $\sim P$ we may begin by introducing P as a premiss; if we are then able to deduce $\sim P$ we use CP to obtain $P \supset \sim P$, whence by interchange we obtain $\sim P$.

Deductions of this kind may be known as *indirect deductions*. One special case of indirect deduction should be mentioned. When we have introduced a premiss P with the object of deducing $\sim P$ and then applying CP one way in which we may obtain $\sim P$ is through the medium of a contradiction. If using P as a premiss we deduce two formulae Q and $\sim Q$ one of which is the contradiction or negation of the other, we may then proceed as follows. First we may use the rule of addition: we apply this to Q and obtain the formula $Q \vee \sim P$. We then apply to this formula and $\sim Q$ the rule MTP and so obtain $\sim P$.

Indirect deductions of this kind correspond to *reductio ad absurdum* arguments in informal reasoning and may be known as *reductio ad absurdum* deductions.

We give now examples of indirect deduction. Example 1 shows that the propositional form $[[p \supset q] \supset p] \supset p$ is a tautology. In example 2 the *reductio ad absurdum* method is used; the deduction shows that the argument form:

$$p \supset q, \ p \supset \sim q; \ \text{therefore} \ \sim p,$$

The Deductive Method

is valid.

Example 1

       Pr × 1
(1) $[p \supset q] \supset p$

       Pr × 2
(2) $\sim p$

      2, add × 3
(3) $\sim p \vee q$
      3, int (VIII) × 4
(4) $p \supset q$
      1, 4, MP × 5
(5) p

     1, (2 ⊢ 5) × 6
(6) $\sim p \supset p$
     6 int (X) × 7
(7) p

    (1 ⊢ 7) × 8

(8) $[[p \supset q] \supset p] \supset p$

Example 2

       Pr × 1
(1) $p \supset q$

       Pr × 2
(2) $p \supset \sim q$

       Pr × 3
(3) p
      1, 3, MP × 4
(4) q
      2, 3, MP × 5
(5) $\sim q$

      4, add × 6
(6) $q \vee \sim p$
      6, 5, MTP × 7
(7) $\sim p$

   1, 2 (1, 2, 3 ⊢ 7)
   × 8
(8) $p \supset \sim p$
      8 int (X) × 9
(9) $\sim p$

SUGGESTED READING

Copi, I. M.: *Symbolic Logic*, chapter iii.
Suppes, P.: *Introduction to Logic* (1957), chapter ii.

Chapter Five

Part I: NORMAL FORMS

1. Introduction. The third method of testing an argument or argument form for validity involves the use of what are called *normal forms*. There are several variants of the method differing from one another according to the kind of normal form that is used. We will describe a method involving the use of *conjunctive* normal forms, and we begin by explaining the sense we are here giving to this term.

A formula is a conjunctive normal form (and is said to be *in* conjunctive normal form) if and only if it is either a disjunction of simple formulae or a conjunction of disjunctions of simple formulae. By *simple formula* is meant here a formula which is, or is the negation of, a truth-functionally atomic formula.

We may illustrate this definition by giving some examples of formulae which are, and other examples of formulae which are not, in conjunctive normal form; in these examples and throughout the rest of this chapter we will denote the negation of an atomic formula a by putting a bar over the formula instead of using the functor $\sim$, i.e. by $\bar{a}$ rather than by $\sim a$. The following formulae are in conjunctive normal form:

$$p \lor q \lor \bar{p}$$
$$[r \lor \bar{q}] . [p \lor \bar{r} \lor s].$$

The first is a disjunction of simple formulae and the second is a conjunction of disjunctions of simple formulae.

Normal Forms

On the other hand p is not in conjunctive normal form, nor is:

$$[\bar{p}.q]\vee\bar{r},$$
$$\text{or } \bar{p}.[q\vee r]$$
$$\text{or } \sim\bar{p}\vee q.$$

2. Use of conjunctive normal forms in testing for validity. The method of testing for validity or invalidity by means of conjunctive normal forms is based on two facts about such forms:

   (i) For every truth-functional formula F there can be found a formula in conjunctive normal form to which F is logically equivalent.

   (ii) If a formula is in conjunctive normal form it is possible to tell at a glance whether or not it is tautologous.

Given these two facts the method can be easily understood. Let us suppose that we have an argument or argument form A which we wish to test for validity. We proceed as follows:

   (a) We find the corresponding implication CI$_\text{A}$.

   (b) We find a formula CI$'_\text{A}$ in conjunctive normal form to which CI$_\text{A}$ is logically equivalent.

   (c) We discover by inspection whether or not CI$'_\text{A}$ is a tautology. If it is then A is truth-functionally valid; otherwise it is truth-functionally invalid.

We will complete our account of the method by explaining (b) and (c) in more detail. (a) has, of course, been dealt with in chapter iii.

   (b) The process of finding a formula F$'$ in conjunctive normal form which is logically equivalent to a given formula F is sometimes referred to as the process of *reducing F to conjunctive normal form*. We use this terminology here.

Normal Forms

We may reduce a formula F to normal form by the following procedure.

(1) Every compound $P \equiv Q$, whether it is part or the whole of F, is replaced by $P \supset Q . Q \supset P$.

(2) Every compound $P \supset Q$ which is not negated is replaced by $\sim P \mathsf{v} Q$.

(3) Every compound $\sim[P \supset Q]$ is replaced by $P . \sim Q$.

(4) Every compound $\sim[P . Q]$ is replaced by $\sim P \mathsf{v} \sim Q$ and every compound $\sim[P \mathsf{v} Q]$ is replaced by $\sim P . \sim Q$. This step is repeated until we have a formula containing no negations of disjunctions and no negations of conjunctions.

(5) Double negations are eliminated and where P is an atomic formula $\sim P$ is written as $\bar{P}$. This may be done all at once at this stage but it will often be more convenient to carry it out piecemeal, partly at least at earlier stages.

We now have a formula F_1 which either is a simple formula or is compounded by . and v out of simple formulae.

(6) If F_1 is not a simple formula or a conjunction of simple formulae we make replacements as often as necessary in accordance with the equivalences:

$$[Q \mathsf{v} R] . P \equiv P . [Q \mathsf{v} R] \equiv [P . Q] \mathsf{v} [P . R]$$
$$[Q . R] \mathsf{v} P \equiv P \mathsf{v} [Q . R] \equiv [P \mathsf{v} Q] . [P \mathsf{v} R].$$

together with the laws of association until we arrive at a formula in normal form.

If F, is itself a simple formula or a conjunction of simple formulae $P_1 . P_2 . \ldots . P_n$, we obtain a normal form by inserting after each P_i the formula $\mathsf{v} P_i$. This obviously justified by the equivalence $P \equiv [P \mathsf{v} P]$.

In the application of this procedure various devices may be used to save unnecessary writing. In the following examples of the reduction of formulae to normal form letters X, Y, Z are used in an obvious way as temporary abbreviations for parts of formulae.

Example 1

  (A) $p \equiv q \cdot \underset{\overline{\text{X}}}{\sim [q \vee r]}$

     $\underset{\overline{\text{Y}}}{[p \supset q] \cdot [q \supset p] \cdot \text{X}}$

     $[\bar{p} \vee q] \cdot \underset{}{[\bar{q} \vee p]} \cdot [\bar{q} \cdot \bar{r}]$
              $\overline{\text{Y}}$

     $[\bar{q} \cdot \bar{r}] \cdot [\bar{p} \vee q] \cdot \text{Y}$

     $\underset{\overline{\text{Z}}}{[\bar{q} \cdot \bar{r} \cdot \bar{p}] \vee [\bar{q} \cdot \bar{r} \cdot q] \cdot \text{Y}}$

     $[Z \vee \bar{q}] \cdot [Z \vee \bar{r}] \cdot [Z \vee q] \cdot \text{Y}$

     $[\bar{q} \vee [\bar{q} \cdot \bar{r} \cdot \bar{p}]] \cdot [\bar{r} \vee [\bar{q} \cdot \bar{r} \cdot \bar{p}]] \cdot [q \vee [\bar{q} \cdot \bar{r} \cdot \bar{p}]] \cdot \text{Y}$

  (A') $[\bar{q} \vee \bar{q}] \cdot [\bar{q} \vee \bar{r}] \cdot [\bar{q} \vee \bar{p}] \cdot [\bar{r} \vee \bar{q}] \cdot [\bar{r} \vee \bar{r}] \cdot [\bar{r} \vee \bar{p}] \cdot [q \vee \bar{q}] \cdot [q \vee \bar{r}]$
        $\cdot [q \vee \bar{p}] \cdot [\bar{q} \vee p].$

A' is a conjunctive normal form equivalent of A. One may check easily that A and A' are logically equivalent. The truth-table for A shows that A is true when p, q and r are all false but false in all other possible cases. By inspection of A' we can see without difficulty that this is true of A' also.

Let us now look at A' from a different point of view. Consider, for example, the last conjunct $[\bar{q} \vee p]$. This must be false in any case in which q is true and p is false. But if the conjunct is false the whole formula is false. Hence we can see that A' is not a tautology. Now look at the conjunct fourth from the end, i.e. $[q \vee \bar{q}]$. This formula is tautologous: there is no case in which it is false, because one or other of the disjuncts is bound to be true, and the truth of one disjunct is sufficient to make the whole conjunct true. The presence of q and $\bar{q}$ as disjuncts in this formula is sufficient to render it a tautology whether or not these are the sole disjuncts. We can see that it is possible to tell at a glance whether a conjunct of a normal

Normal Forms

form formula is or is not a tautology: if it contains among its disjuncts $\bar{P}$ and P it is a tautology; otherwise it is not. Now if all the conjuncts are tautologies the whole formula is a tautology, but if any conjunct is not a tautology the whole formula is not a tautology. Since $[\bar{q}vp]$ is a conjunct of A′ but is not a tautology A′ itself is not a tautology; hence the equivalent formula A is not a tautology either.

Since by reducing a formula to normal form we are able to find out whether or not it is a tautology we can, of course, do this for the corresponding implication for an argument and so discover whether or not the argument is valid. For example to test whether the following argument form:

(B) $\sim[q\equiv r]$; therefore $qv\sim r$

is valid we may first form:

(CI$_B$) $\sim[q\equiv r]\supset[qv\sim r]$

and then use the normal form method to discover whether or not CI$_B$ is a tautology.

A reduction of CI$_B$ to an equivalent normal form CI′$_B$ is now shown:

$$
\begin{array}{l}
(\text{CI}_B) \ \sim[q\equiv r]\supset \underline{qv\sim r} \\
\phantom{(\text{CI}_B) \ \sim[q\equiv r]\supset} \text{X} \\
\quad \sim[[q\supset r].[r\supset q]]\supset\text{X} \\
\quad \sim[[\bar{q}vr].[\bar{r}vq]]\supset\text{X} \\
\quad [[\bar{q}vr].[\bar{r}vq]]v\text{X} \\
\quad \text{X}v[[\bar{q}vr].[\bar{r}vq]] \\
\quad [\text{X}v\bar{q}vr].[\text{X}v\bar{r}vq] \\
(\text{CI}'_B) \ [qv\bar{r}v\bar{q}vr].[qv\bar{r}v\bar{r}vq].
\end{array}
$$

The second conjunct in CI′$_B$ is not a tautology. Therefore CI′$_B$ is not a tautology and so CI$_B$ is not tautologous and B is not a valid argument form. Consider, however, the argument form

(C) $\sim[q\equiv r]$; therefore $\sim[q.r]$.

100

Normal Forms

We form:

(CI_C) $\sim[q\equiv r]\supset\sim[q.r]$

and reduce it to a normal form CI'_C.

$$(CI_C)\quad \sim[q\equiv r]\supset\underline{\sim[q.r]}$$
$$ X$$
$$\sim[[q\supset r].[r\supset q]]\supset X$$
$$[[q\supset r].[r\supset q]]\vee X$$
$$[[\bar{q}\vee r].[\bar{r}\vee q]]\vee\underline{[\bar{q}\vee\bar{r}]}$$
$$\phantom{[[\bar{q}\vee r].[\bar{r}\vee q]]\vee} Y$$
$$Y\vee[[\bar{q}\vee r].[\bar{r}\vee q]]$$
$$[Y\vee\bar{q}\vee r].[Y\vee\bar{r}\vee q].$$
$$(CI'_C)\quad [\bar{q}\vee\bar{r}\vee\bar{q}\vee r].[\bar{q}\vee\bar{r}\vee\bar{r}\vee q].$$

Each conjunct in CI'_C is a tautology; so CI'_C and CI_C are tautologies and C is a valid argument form.

3. Theoretical use of normal forms. The method of normal forms is often in practice a rather tedious way of discovering whether or not an argument is valid. However, for theoretical purposes normal forms are frequently highly useful. For example, if one wants to prove some general statement S to be true of all possible truth-functional formulae this may seem at first a formidable task. However, it may be possible to show that S holds for all normal form formulae. If this is so and it can also be shown (i) that if S holds for a given formula it holds for any formula equivalent to the given one and (ii) that every formula is equivalent to some formula in normal form then of course S must hold for all formulae.

As an example we will show in outline at least how it may be proved that the deductive system which we studied in the last chapter is complete in the sense that every tautology may be deduced within it. From this, of course, it will follow that every truth-functionally valid argument or argument form may be shown to be valid by a deduction within our system.

Consider any formula F. In a full proof our first step would be to show that, whatever F is, a formula F′ may be found such that (i) F′ is in normal form and (ii) F′ is logically equivalent to F. We set out earlier a six-stage procedure for reducing a formula to normal form; we take it for granted here that this procedure is universally effective; in the complete proof of which this is an outline, this point, of course, would have to be proved.

Next we show how, when we have found for F a logically equivalent normal form F′, we are able if F is a tautology to show that this is so by means of a deduction within the deductive system of the last chapter.

If F is a tautology the logically equivalent F′ must be a tautology also. Hence F′ will contain in every conjunct two mutually contradictory disjuncts P and $\bar{P}$. We now show how F may be deduced within our deductive system in every such case. For each conjunct in F′ we first deduce in our system a formula αv where α and $\bar{\alpha}$ are two mutually contradictory disjuncts of that conjunct. This deduction is done thus:

$$
\begin{array}{ll}
 & \text{Pr} \times 1 \\
(1) & \sim\alpha \\
 & 1 \text{ add} \times 2 \\
(2) & \sim\alpha v\alpha \\
 & 2 \text{ int (II)} \times 3 \\
(3) & \alpha v \sim\alpha \\
 & 3, 1, \text{MTP} \times 4 \\
(4) & \sim\alpha
\end{array}
$$

$$
\begin{array}{ll}
 & (1 \vdash 4) \times 5 \\
(5) & \sim\alpha \supset \sim\alpha \\
 & 5 \text{ int (VIII)} \times 6 \\
(6) & \alpha v \sim\alpha
\end{array}
$$

Applicability and Limitations

Having thus obtained αvᾱ for a particular disjunction D we use the rule of addition to obtain a disjunction D′ which has the same disjuncts as D. We then use the rule of interchange in respect of the association and commutative laws (III, II) to rearrange the order of the disjuncts as often as is necessary until eventually we obtain D. In a similar way we obtain each conjunct of F′. We now apply the rule of conjunction repeatedly until we obtain F′ itself. Our next task is to retrace in the opposite direction the informal reductive procedure by which we got to F′ from F. Since each step in this procedure was by substitution of equivalents in accordance with one of the equivalences of our system it is evident that in a formal deduction we may reverse the process and, using the rule of interchange at each step in respect of the equivalence used at the corresponding step in the reduction, obtain a deduction which begins with F′ and ends with F. It has already been shown, however, that we are able to obtain a deduction of F′ (with no undischarged premiss). Putting the two together we have a deduction which shows that F is a tautology. All this of course holds for any tautology F. Thus we have shown how with the help of normal forms it is possible to prove an interesting completeness theorem.

Part II: APPLICABILITY AND LIMITATIONS OF TRUTH-FUNCTIONAL LOGIC

4. General remarks. In deductive reasoning we start with a certain piece or with certain pieces of information and go on to obtain other information. We do this by means of thought alone without observation or any kind of empirical inquiry. Or, alternatively, we convince ourselves that if we were given certain information we could

103

obtain therefrom other information by means of thought alone. In formal logic we study ways in which deductive reasoning may be done correctly. A system of formal logic may be regarded as an instrument which enables us to do certain kinds of reasoning correctly or to check whether such reasoning has been done correctly in particular cases. Any such system is applicable directly only to reasoning involving arguments in which the propositions are of some standard type. Let us say that truth-functional logic is applicable directly only to arguments which contain and depend on the symbols which we have called truth-functors.

It is applicable indirectly to arguments which are intimately related to these in a way which will be described presently. We must distinguish, however, between fields of reasoning in which truth-functional logic is by itself a sufficient instrument and others where though not sufficient it is necessary or at any rate highly useful. As an example of the latter we may mention quantificational arguments which depend on notions corresponding to *all*, *some*, *any* and other related expressions. Many quantificational arguments contain an important truth-functional element and in the normal contemporary treatment the principles of truth-functional logic are used along with others specifically quantificational.[1] There are indeed perhaps few branches of logic in which truth-functional logic is not involved either explicitly or implicitly. Our present concern, however, is with the application, direct or indirect, of truth-functional logic where its methods alone are sufficient. We should mention first perhaps that truth-functional logic has possible uses of a technical kind. It can be used for example in the solution of problems concerned with the design of various kinds of electrical circuit, though such problems in practice tend to be sub-

[1] Quantificational logic is the subject of two monographs in this series.

mitted to the methods of the related discipline of Boolean algebra.[1] However, we do not propose to discuss here the possibility of technical applications. We will consider instead a question of more general interest.

Truth-functional logic is treated in many, though not in all, text-books of formal logic as being applicable, more or less without restriction, to such arguments of ordinary discourse as depend for their force on the words: *not*, *if*, *and*, *or* and their equivalents. On the other hand there are many people who, mainly as a result of doubts concerning the relationship of *if* and ⊃, would question the legitimacy of applying truth-functional logic to these arguments except perhaps in a few extreme and rather trivial cases. To what extent then is truth-functional logic in fact applicable to arguments of the kind described? This is a question about the applicability of truth-functional logic which it would be wrong to ignore. For on the answer to it depends, not indeed the existence of the subject, but its importance in relation to general education: is it on the one hand a subject of quite wide interest and significance or, on the other hand, is it, or is it akin to, one of those branches of mathematics which are the concern of specialists only?

We are not able to deal here with every aspect of this question. There is one major problem as well as a number of relatively unimportant minor ones. We will attempt to isolate the major problem and will then confine our attention to it.

5. Special terminology. The broad field with which our question is concerned is that of what we shall call *ordinary discourse propositional arguments*. We will define presently two groups of arguments: group I arguments and group II arguments, and we may say at once that what we mean by an ordinary discourse propositional argument

[1] Boolean algebra is the subject of another monograph.

105

Applicability and Limitations

is any argument which belongs to either group. Before defining group I we must explain the expression *formal truth-functional argument*. A formal truth-functional argument is one which depends for its force or plausibility entirely on the truth-functors $\sim$, $\supset$, ., v, $\equiv$. All the arguments which we have used as examples in chapters iii and iv have, of course, been formal truth-functional arguments in this sense. Now to every formal truth-functional argument Z_t there corresponds an argument Z_s which can be constructed by making replacements throughout Z_t as follows: $\sim p$ is replaced by *not p*, $p \supset q$ by *if p then q*, $p.q$ by *p and q*, pvq by *p or q* and $p \equiv q$ by *p if and only if q*. The *or* which replaces v must always be understood in the inclusive sense and the *and* which replaces . must have no temporal significance. Any argument Z_s which can be formed in this way from a formal truth-functional argument Z_t will be known as a *stereotype argument*. Z_s and Z_t will be referred to as *corresponding arguments*: it can be seen that to every formal truth-functional argument Z_t there corresponds a single stereotype argument Z_s and that to every stereotype argument there corresponds a single formal truth-functional argument. The propositions which make up a formal truth-functional argument will be referred to as formal truth-functional propositions and those which make up a stereotype argument will be known as stereotype propositions. Each stereotype proposition F_s, of course, corresponds to a single formal truth-functional proposition F_t and conversely. An argument Z (with conjunction of premisses P and conclusion C) will be said to have a *stereotype counterpart* if and only if there is a stereotype argument Z_s (with conjunction of premisses P_s and conclusion C_s) such that P_s has the same meaning as P and C_s has the same meaning as C. It follows that every stereotype argument has a stereotype counterpart, namely itself. Of non-stereotype arguments some have stereotype counterparts and others not. We are now in a position to define our two groups of arguments:

Group I arguments are those which have stereotype counterparts.

Group II arguments are arguments not belonging to group I of which the force depends entirely on some at least of the expressions *not, if, and, or, if and only if* and their equivalents.

A complete discussion of the applicability of truth-functional logic to ordinary discourse propositional arguments would involve consideration of some rather complex questions, for example, about whether arguments of certain types have or have not stereotype counterparts. Since such questions cannot be dealt with here we will confine our attention to group I. It will be argued that truth-functional logic can be used in determining the validity or invalidity of any argument belonging to this group. If this is correct truth-functional logic has a useful and important application to ordinary discourse.

6. Argument begins in support of the proposition that truth-functional logic is applicable to all group I arguments; lemma 1 and lemma 2. Truth-functional logic is applicable indirectly in a worthwhile way to all group I arguments if every group I argument Z satisfies the condition that there is a formal truth-functional argument which is valid if and only if Z is valid. Let us call this condition C. Since every group I argument which is not itself stereotype has a stereotype counterpart it is evident that if condition C is satisfied by all stereotype arguments it is satisfied by all group I arguments. It will now be argued that all stereotype arguments do in fact satisfy condition C.

We begin by defining the notions of *derivability* and *interderivability*. A proposition Y is derivable from a pro-

H   107

Applicability and Limitations

position X if Y can be inferred from X alone, i.e. if we can know that Y must be true if X is true no matter what other propositions we know to be true or false. X and Y are interderivable if Y is derivable from X and X is derivable from Y. We will use $X \to Y$ to mean that Y is derivable from X and $X \leftrightarrow Y$ to mean that X and Y are interderivable.

Our contention will be that any stereotype argument Z_s satisfies condition C simply by virtue of the fact that Z_s is valid if and only if the single corresponding formal truth-functional argument Z_t is also valid. Let P_s, P_t be the conjunctions of premisses of Z_s and Z_t respectively and let C_s and C_t be the respective conclusions. It is not difficult to see that Z_s is valid if and only if Z_t is valid, provided that the following condition D is satisfied: (Condition D): $P_s \leftrightarrow P_t$ and $C_s \leftrightarrow C_t$. We will now proceed to argue that in fact any stereotype proposition F_s is interderivable with the formal truth-functional proposition F_t to which it corresponds.

Let F_s be any stereotype proposition and let F_t be the formal truth-functional proposition corresponding to F_s. Our argument that $F_s \leftrightarrow F_t$ is based on two lemmas which refer to certain *interderivability postulates*. These are: (i) $\sim p \leftrightarrow not\ p$, (ii) $p \supset q \leftrightarrow if\ p\ then\ q$, (iii) $p.q \leftrightarrow p\ and\ q$, (iv) $p \lor q \leftrightarrow p\ or\ q$, (v) $p \equiv q \leftrightarrow p\ if\ and\ only\ if\ q$. The lemmas are: *lemma 1*, if the interderivability postulates are true then $F_s \leftrightarrow F_t$; *lemma 2*, the interderivability postulates are true.

The argument in support of lemma 1, though not entirely obvious, is not difficult. For reasons of space it cannot be given here but must be taken for granted; fortunately few people are likely to be inclined to dispute the truth of this lemma. In the case of lemma 2 postulates (i) and (iii) are obvious and the main difficulties which might arise in connexion with (iv) and (v) are dependent on difficulties connected with (ii). We will therefore assume

Applicability and Limitations

that all the postulates are true if (ii) is true. Our argument in support of lemma 2 will consist solely of an argument in support of the interderivability of $p \supset q$ and *if p then q*. The relationship between *if* and $\supset$ is in fact the crux of the whole topic of applicability. Section 7 will be devoted to a discussion of this relationship and will include in its early part a certain amount of material not strictly necessary to our argument in support of lemma 2 which itself extends to the last page of the book.

7. Relation between *if* and $\supset$: preliminary discussion. Many people regard *if* as being essentially non-truth-functional in its use and are bewildered by the suggestion that it can be defined by the truth-table for $\supset$ or that an *if-then* statement is true always and only when the corresponding $\supset$ statement is true. The same sort of puzzlement does not arise in connexion, for example, with *or* and ∨ or with *and* and . . Although there may be doubt about the interderivability in general of *or* and ∨ the point of assigning to *or* the truth-table for ∨ is easily understood; for it is recognized that *or*, even if it may not always be used truth-functionally, is at least sometimes used truth-functionally in the way suggested. But with *if*, on the other hand, the position is quite different. An *if-then* statement is regarded as essentially asserting a connexion of some kind between antecedent and consequent; as we shall see later this appears incompatible with truth-functionality; consequently *if* unlike *or* and *and* is thought not to have a truth-functional use at all. Now whatever the truth may be about the relation between *if* and $\supset$ I think that at least this extreme view that there can be no natural truth-functional use of *if* is very hard to defend although one can quite see how people come to adopt it. I shall begin by giving an example of a situation in which it seems quite natural to interpret *if* truth-functionally in accordance with the truth-table for $\supset$.

109

Applicability and Limitations

Our example relating to a truth-functional use of *if* will be preceded for the sake of comparison by an analogous one concerned with *and*. Let us suppose that it is a condition for appointment to a certain post that the candidate should be over twenty-one and a graduate. If a certain candidate Robinson is to be regarded as satisfying this condition it must be the case that:

(i) Robinson is over twenty-one and Robinson is a graduate. It will be convenient to use abbreviations; we shall put *T* for *Robinson is over twenty-one* and *G* for *Robinson is a graduate*. (i) thus becomes:

(i) *T* and *G*.

We shall call (i), in the present context, the qualification statement. Now there are four possibilities about truth-values for the propositions *T* and *G*. In one of the possible cases, namely *T* and *G* both true, Robinson satisfies the condition; in the other three cases he fails to satisfy it. But also in the first case, *T* and *G* both true, the qualification statement (i) is true and in the other three cases the qualification statement is false. Thus the qualification statement is true if and only if Robinson satisfies the condition. We may set all this out in a table:

Table I

T	G	Condition satisfied	Truth-value of qualification statement (i)
1	1	Yes	1
1	0	No	0
0	1	No	0
0	0	No	0

On the other hand there might be a different condition of appointment namely that if the candidate is over twenty-

110

one he should be a graduate. If Robinson is to be regarded as satisfying this condition it must be the case that

(ii) If T then G.

(ii) is in this case the qualification statement. Again there are four possibilities about truth-values for T and G. In the first case where T and G are both true clearly Robinson satisfies the condition; in the second case, where T is true but G is false, he fails to satisfy the condition; in the third and fourth cases, in both of which T is false, he satisfies the condition simply by not being over twenty-one; for clearly he either satisfies or fails to satisfy it and he cannot fail to satisfy it unless he is over twenty-one. Thus Robinson fails to satisfy the condition in the second case, where T is true and G is false, but satisfies it in all the other possible cases. Now what about the truth-value of the qualification statement? Surely, just as in the previous example, we must say that this statement is true if Robinson satisfies the condition and false if he fails to satisfy it. But, if so, the use of *if* in (ii) is truth-functional. For all we need to know about T and G in order to know whether the condition is satisfied or not are their truth-values. Further, the truth-table for *if* in this case will be exactly the same as the truth-table for $\supset$. Again a table brings these points out clearly:

Table II

T	G	Condition satisfied	Truth-value of qualification statement (ii)
1	1	Yes	1
1	0	No	0
0	1	Yes	1
0	0	Yes	1

Applicability and Limitations

8. Argument against the interderivability of *if* and $\supset$. The paradoxes. The example we have just been using does, I think, show that there are some occasions at least on which it would be natural and not just perverse to use *if* in a purely truth-functional sense. However, it by no means follows from this that in general *if* is used in a way in which *if* and $\supset$ are interderivable, and in fact very plausible arguments can be adduced in support of the contrary view that, at least as a general rule, *if* and $\supset$ are not interderivable. These arguments have their focus in what have been called the paradoxes of implication or of material implication and these we will now explain.

It can be seen from the truth-table for $\supset$ that the proposition $p \supset q$, which is said to assert the material implication by p of q, is false when p is true and q is false but true for all other possible values of p and q. Another way of saying exactly the same thing is to say that $p \supset q$ is true if p is false or q is true but otherwise it is false. In other words if a certain proposition p is false then whatever proposition q may be the proposition $p \supset q$ is true; and again if q is a true proposition then whatever proposition p may be $p \supset q$ is true. Now if we replace $\supset$ by *if-then* the sentence we have just written will read ' . . . if a certain proposition p is false then whatever proposition q may be the proposition *if p then q* is true; and again if q is a true proposition then whatever proposition p may be the proposition *if p then q* is true'. This sentence now expresses the so-called *paradoxes of material implication*. They may also be put in rather different language: if $p \supset q$ is taken to assert that p implies q the paradoxes are that a false proposition implies any proposition and a true proposition is implied by any proposition. One or two examples will make it clear why these assertions are described as paradoxical. The proposition *Daniel Defoe lived in the fifteenth century* is false; hence according to those assertions if we take any other proposition whatever, say for example,

112

Applicability and Limitations

Men will land on the moon before 1970 the compound proposition:

(iii) If Daniel Defoe lived in the fifteenth century then men will land on the moon before 1970

is true. Again the proposition *Daniel Defoe lived in the fifteenth century* is said to imply the proposition *Men will land on the moon before 1970*. These statements are paradoxical because as many people understand the word *if* the proposition (iii) cannot be true unless there is some connexion between the proposition *Daniel Defoe lived in the fifteenth century* and the proposition about men landing on the moon; there is obviously, it seems, no connexion and yet if the interderivability of *if* and $\supset$ is allowed the proposition (iii) is true.

To take another example, relevant this time to the second paradoxical assertion, the proposition $2+2=4$ is true. It follows, according to the second assertion, that the proposition:

(iv) If Cicero was a poet $2+2=4$

is true. But again there seems to be no connexion between the proposition that Cicero was a poet and the proposition that $2+2=4$; for this reason it seems very queer to call (iv) a true proposition.

These paradoxes are the basis of the main argument against the interderivability of *if* and $\supset$. They are also the basis of the main argument against the *synonymity* of *if* and $\supset$. This distinction must be explained before we go further. Propositions X and Y, as we have seen, are interderivable if and only if each of them can be inferred from the other alone. On the other hand X and Y are synonymous propositions if and only if they have the same meaning— or perhaps strictly we should say, if every sentence which expresses the proposition X has the same meaning as every sentence which expresses the proposition Y.

113

Applicability aud Limitations

Now let us formulate briefly the arguments, based on the paradoxes, against interderivability and synonymity respectively. *Argument against interderivability*. There is some kind of connexion the existence of which between *p* and *q* is a necessary condition of the truth of *if p then q* but is not a necessary condition of the truth of $p \supset q$. Therefore it is possible for $p \supset q$ to be true when *if p then q* is false; and so *if* and $\supset$ are not interderivable. *Argument against synonymity*. There is some kind of connexion which is part of the meaning of *if p then q* but is not part of the meaning of $p \supset q$. Therefore *if* and $\supset$ are not synonymous.

In the next section we put forward an argument in support of interederivability. Part of our contention will be that the only kind of connexion which can plausibly be held to be a necessary condition of the truth of *if p then q* is also a necessary condition of the truth of $p \supset q$. If this is correct, the argument stated above against interderivability is destroyed; for its premiss, despite the paradoxes, must be false. However, in case there should be misunderstanding it should be said that our argument for interderivability will not, by itself at least, destroy the argument against synonymity. Interderivability is not in general incompatible with non-synonymity. Consider the following propositions:

(A) The winner was the tallest and James was the winner.

(B) The winner was the tallest and James was the tallest.

These two propositions are interderivable but they are not synonymous.

Before we leave the subject of the paradoxes it ought I think to be pointed out that they are rather misleadingly described as paradoxes of *material implication*. To describe them thus is to suggest that the paradoxes concern the

114

functor ⊃ in itself, and that paradox is involved, not
merely in calling (iii) and (iv) true propositions, but even
in calling the following proposition (v) and (vi) true:

> (v) Daniel Defoe lived in the fifteenth century
> ⊃ Men will land on the moon before 1970.
> (vi) Cicero was a poet ⊃ 2+2=4.

This would be a mistake. The functor ⊃ has no meaning except what is defined by the relevant truth-table.
From the truth-table for ⊃ together with the falsity of
the antecedent of (v) and the truth of the consequent of
(vi) the truth of (v) and (vi) necessarily follows and there
is no paradox whatsoever. What paradox there is arises
only from taking ⊃ and *if-then* to be interderivable or
synonymous.

**9. Completion of argument in support of lemma 2:
argument for interderivability.** To show that *if*
and ⊃ are interderivable we have to establish both the
derivability of ⊃ from *if* and the derivability of *if* from ⊃ ;
i.e. we have to establish both that (i) $p \supset q$ is derivable
from *if p and q* and that (ii) *if p then q* is derivable from
$p \supset q$. The derivability of ⊃ from *if* hardly needs any
proof; for everyone would agree that when *if p then q* is
true it cannot be the case that *p* is true and *q* is false. But
if it is not the case that *p* is true and *q* is false we know
from the truth-table for ⊃ that $p \supset q$ is true. But that *if p
then q* is derivable for $p \supset q$ certainly does require to be
argued and to this we now proceed.

We said in section 8 that the existence of a certain connexion between *p* and *q* is commonly believed to be a
necessary condition of the truth of *if p then q*. We have
been regrettably vague about this connexion and we must
now inquire what its character is. At first thought we
might perhaps be inclined to say that the connexion is
simply that *q* is derivable from *p*. This works all right for

some cases; for example if *p* is *No Frenchmen were saved* and *q* is *No one who was saved was a Frenchman* then in the *if-then* proposition:

(i) If no Frenchmen were saved no one who was saved was a Frenchman,

q is certainly derivable from *p*. However, the suggestion in its present form does not always work. For if *p* is for example *Smith is taller than Jones* and *q* is *Smith is taller than Robinson* then in the *if-then* proposition:

(ii) If Smith is taller than Jones Smith is taller than Robinson,

q certainly cannot be derived from *p* (i.e. inferred from *p* alone). Yet (ii) is a very everyday sort of *if-then* proposition which may well be true and which any theory must be able to account for. Let us for the moment confine our attention to (ii), referring to the antecedent as *p* and to the consequent as *q*. Although *q* cannot be inferred from *p* alone it can be inferred from *p* together with the proposition,

(iii) Jones is at least as tall as Robinson.

Now it might perhaps be thought that (ii) can be true if and only if (iii) is true and that the connexion which is a condition of the truth of (ii) is simply the relationship asserted in (iii) or at least is based on this relationship. However, this cannot be right. Certainly (ii) will be true if (iii) is true but not *only* if (iii) is true. It is not difficult to think of an example to demonstrate that (ii) can be true even if (iii) is false. For example, if the following three propositions were true:

(iv) Every member taller than Jones is red-haired;
(v) Every red-haired member is taller than Robinson;

(vi) Smith is a member;

the proposition

(ii) If Smith is taller than Jones Smith is taller than Robinson

would also be true. But (iv), (v) and (vi) could quite well all be true even though (iii) were false; for example if Smith and Jones are the only members, Smith is red-haired and Jones not, and Smith, Robinson and Jones are respectively 6 ft., 5 ft. and 4 ft. tall then (iv), (v) and (vi) are true but (iii) is false. Hence in this case (ii) is true though (iii) is false. In other words the truth of (iii) though a sufficient is not a necessary condition of the truth of (ii). But the connexion about which we are inquiring has to be a necessary condition of the truth of *if p then q*.

From this suggestion we may pass on to one that is more promising. Perhaps the condition for the truth of *if p then q* is not that there should be a specified true proposition (as e.g. (iii) in our last example) from which together with *p q* is inferrible but simply that there should be *some* true proposition or propositions from which together with *p q* is inferrible. We may formulate this condition as follows:

(Condition E.) There is a set S of true propositions such that *q* is inferrible from *p* together with S.

The effect of this condition may be illustrated by reference to our previous example. If (iii) is a true proposition then there is a set S of true propositions, namely the set consisting of (iii) itself, such that *q* is inferrible from *p* together with S. Thus condition E is satisfied. On the other hand if (iii) is not true the condition may still be satisfied; for example if propositions (iv), (v) and (vi) are true there is a set S of true propositions, namely the set consisting of these three propositions, such that *q* is inferrible from *p* together with S. Thus condition E is again satisfied. We can imagine of course many different ways besides these two that we have mentioned in which condition E might be satisfied for this particular example.

It seems to me that, whatever propositions p and q may be, condition E is a sufficient and necessary condition of the truth of *if p then q*. I think it is sufficient because I think that in any case in which we believed that a set S existed as specified we should be prepared to assert *if p then q* and I think it is necessary because I think that in any case in which we believed that no such set existed we should be prepared to deny *if p then q*. It is to be understood of course that in a case in which q is inferrible from p alone the condition E is satisfied in that S may then be taken to be any set of true propositions whatever. If we may refer to an earlier example in which p was the proposition, *No Frenchmen were saved* and q was the proposition, *No one who was saved was a Frenchman* we see that in the case of the proposition

(i) If no Frenchmen were saved then no one who was saved was a Frenchman,

the condition E is satisfied because since q is inferrible from p alone it is inferrible from any set S of true propositions together with p.

We take it then that condition E is a sufficient and necessary condition of the truth of *if p then q*; and we now return to the question of the relationship between *if p then q* and $p \supset q$. Now q can certainly be inferred from p and $p \supset q$. Accordingly if $p \supset q$ is true there is a set S of true propositions, namely the set consisting solely of the proposition $p \supset q$, such that q is inferrible from p together with S. It follows that if $p \supset q$ is true condition E is satisfied. But if condition E is satisfied *if p then q* is true. Consequently, if we are able to know that the proposition $p \supset q$ is true we are able to know that the proposition *if p then q* is true also whatever other truths or falsehoods there may be. That is to say *if p then q* is derivable from $p \supset q$. But we have already seen that $p \supset q$ is derivable

118

Applicability and Limitations

from *if p then q*. It follows that the propositions *if p then q* and $p \supset q$ are interderivable.

This completes the argument in support of lemma 2. It follows from lemmas 1 and 2 that any stereotype proposition F_s is interderivable with the corresponding truth-functional proposition F_t. It is immediately obvious that it follows from this that condition D is satisfied for any stereotype argument Z_s, and if this is so truth-functional logic is applicable to all arguments of group I.

SUGGESTED READING

PART 1

Hilbert, D., & Ackermann, W.: *Principles of Mathematical Logic*, chapter i, sections 3 and 4.

Copi, I. M.: *Symbolic Logic*, chapter ii.

PART 2

Technical application

Kemeny, J. G., Snell, J. L., & Thompson, G. L.: *Introduction to Finite Mathematics*, chapter i.

Relation to ordinary discourse

Strawson, P. F.: *Introduction to Logical Theory* (1952), chapter iii.

Blanshard, B.: *The Nature of Thought* (1939), chapter xxix.

Russell, L. J.: *Formal Logic and Ordinary Language* (Analysis, December 1960).

Quine, W. V.: *Methods of Logic* (1952), section 3.

Von Wright, G. H.: *On Conditionals* (in Logical Studies (1957)).

Prior, A. N.: *Formal Logic* (1955), chapter i.

LIST OF ABBREVIATIONS

Add	Rule of addition	Int	Rule of Interchange
CI$_F$	Implication correspond-	Simp	Rule of Simplification
	ing to F	MP	*Modus ponens*
Conj	Rule of Conjunction	MT	*Modus tollens*
CP	Rule of conditional proof	MTP	*Modus tollendo ponens*
ES	Explanatory Supplement	Pr	Premiss rule
HS	Rule of hypothetical		
	syllogism		

INDEX OF DEFINITIONS

References are to pages on which symbols or terms are defined or otherwise explained.

I. SYMBOLS

(*i*) *Symbols used*

Index of Definitions

II. TERMS

121

Index of Definitions